NEWS

NASA

NATIONAL AERONAUTICS AND SPACE ADMINISTRATION
WASHINGTON, D.C. 20546

TELS. WO 2-4155
WO 3-6925

FOR RELEASE: SUNDAY
February 23, 1969

RELEASE NO: 69-29

PRESS KIT

PROJECT: APOLLO 9

I0030982

contents

GENERAL RELEASE---1-11
APOLLO 9 COUNTDOWN--12-13
SEQUENCE OF EVENTS---14-16
MISSION OBJECTIVES--17
MISSION TRAJECTORY AND MANEUVER DESCRIPTION--------------------18
    Launch--18
    Transposition and Docking----------------------------------18
    LM Ejection and Separation---------------------------------18
    S-IVB Restarts---18
    Service Propulsion System (SPS) Burn No. 1-----------------19
    Docked SPS Burn No. 2--------------------------------------19
    Docked SPS Burn No. 3--------------------------------------19
    Docked SPS Burn No. 4--------------------------------------19
    LM Systems Checkout and Power-up---------------------------19
    Docked LM Descent Engine Burn------------------------------19
    Docked SPS Burn No. 5--------------------------------------20
    Extravehicular Activity------------------------------------20
    CSM RCS Separation Burn------------------------------------20
    LM Descent Engine Phasing Burn-----------------------------20
    LM Descent Engine Insertion Burn---------------------------21
    CSM Backup Maneuvers---------------------------------------21
    LM RCS Concentric Sequence Burn (CSI)----------------------21
    LM Ascent Engine Circularization Burn (CDH)----------------21
    LM RCS Terminal Phase Initialization Burn------------------21
    LM Ascent Engine Long-Duration Burn------------------------22
    SPS Burn No. 6---22
    SPS Burn No. 7---22
    SPS Burn No. 8---22
    Entry--22

-more-

2/14/69

Cover: This view of the Apollo 9 command and service module was photographed from lunar module 'Spider' on fifth day of the mission.

Image Credit: NASA

Published by Books Express Publishing
Copyright © Books Express, 2012
ISBN 978-1-78039-858-7

Books Express publications are available from all good retail and online booksellers. For publishing proposals and direct ordering please contact us at: info@books-express.com

Contents Continued

APOLLO 9 MISSION EVENTS--23-25
FLIGHT PLAN SUMMARY---26-28
APOLLO 9 ALTERNATE MISSIONS---------------------------------------29-30
ABORT MODES---31
APOLLO 9 GO/NO-GO DECISION POINTS---------------------------------32
RECOVERY OPERATIONS---33
PHOTOGRAPHIC EQUIPMENT--34
APOLLO 9 EXPERIMENT (Multispectral Photography)-------------------35-36
APOLLO 9 ONBOARD TELEVISION---------------------------------------37
COMMAND AND SERVICE MODULE STRUCTURE, SYSTEMS---------------------38
  CSM Systems---39-41
LUNAR MODULE STRUCTURES, SYSTEMS----------------------------------42-43
  LM-3 Spacecraft Systems-- 44-47
FACT SHEET, SA-504--48
LAUNCH VEHICLE--49
  Saturn V--49-53
  Launch Vehicle Instrumentation and Communication---------------53
  Vehicle Weights During Flight----------------------------------54-55
  S-IVB Restarts---55-56
  Differences in Apollo 8 & Apollo 9 Launch Vehicles-------------57-58
  Launch Vehicle Sequence of Events------------------------------58-60
  Launch Vehicle Key Events--------------------------------------61
LAUNCH FACILITIES---62
  Kennedy Space Center-Launch Complex----------------------------62-64
  KSC Launch Complex 39--65-66
  The Vehicle Assembly Building----------------------------------66-67
  The Launch Control Center--------------------------------------67-68
  Mobile Launcher--68-69
  The Transporter--69-71
  Mobile Service Structure---------------------------------------71
  Water Deluge System--71-72
  Flame Trench and Deflector-------------------------------------72
  Pad Areas--72-73
MISSION CONTROL CENTER--74-75
MANNED SPACE FLIGHT NETWORK---------------------------------------76
  Network Configuration for Apollo 9-----------------------------77
  NASA Communications Network - Goddard--------------------------78-80
SHIPS AND AIRCRAFT NETWORK SUPPORT - Apollo 9---------------------81
  Apollo Instrumentation Ships (AIS)-----------------------------81
  Apollo Range Instrumentation Aircraft (ARIA)-------------------82
APOLLO 9 CREW---83
  Crew Training--83-84
  Crew Life Support Equipment------------------------------------84-85
  Apollo 9 Crew Meals--86-90
  Personal Hygiene---91
  Survival Gear--91-92
  Biomedical Inflight Monitoring---------------------------------92
  Crew Launch-Day Timeline---------------------------------------92-93
  Rest-Work Cycles---93
  Crew Biographies---94-98

Contents Continued

APOLLO PROGRAM MANAGEMENT/CONTRACTORS-----------------------99
  Apollo/Saturn Officials---------------------------------99-101
  Major Apollo/Saturn V Contractors----------------------102-103
APOLLO 9 GLOSSARY--104-107
APOLLO 9 ACRONYMS AND ABBREVIATIONS-----------------------108-109
CONVERSION FACTORS--110-111

-end-

FOR RELEASE: SUNDAY
February 23, 1969

RELEASE NO: 69-29

APOLLO 9 CARRIES LUNAR MODULE

Apollo 9, scheduled for launch at 11 a.m. EST, Feb. 28, from the National Aeronautics and Space Administration's Kennedy Space Center Launch Complex 39A, is the first manned flight of the Apollo spacecraft lunar module (LM).

The Earth-orbital mission will include extensive performance tests of the lunar module, a rendezvous of the lunar module with the command and service modules, and two hours of extravehicular activity by the lunar module pilot.

To the maximum extent possible, the rendezvous in Earth orbit will resemble the type of rendezvous that will take place in lunar orbit following a lunar landing. Rendezvous and docking of the lunar module with the command and service modules, extensive testing of the lunar module engines and other systems, and extravehicular activity are among the mission's objectives.

-more-   2/14/69

Apollo 9 crewmen are Spacecraft Commander James A. McDivitt, Command Module Pilot David R. Scott and Lunar Module Pilot Russell Schweickart. The mission will be the second space flight for McDivitt (Gemini 4) and Scott (Gemini 8), and the first for Schweickart.

Backup crew is comprised of Spacecraft Commander Charles Conrad, Jr., Command Module Pilot Richard F. Gordon and Lunar Module Pilot Alan L. Bean.

Mission events have been arranged in a work-day basis in what is perhaps the most ambitious NASA manned space mission to date. The first five days are packed with lunar module engine tests and systems checkouts, burns of the service module's 20,500-pound-thrust engine while the command/service module and lunar module are docked, the lunar module pilot's hand-over-hand transfer in space from the lunar module to the command module and back again, and rendezvous. The remainder of the 10-day, open-ended mission will be at a more leisurely pace.

The first day's mission activities revolve around docking the command module to the lunar module still attached to the Saturn V launch vehicle S-IVB third stage. When docking is complete and the tunnel joint between the two spacecraft is rigid, the entire Apollo spacecraft will be spring-ejected from the S-IVB. Maneuvering more than 2,000 feet away from the S-IVB, the Apollo 9 crew will observe the first of two restarts of the S-IVB's J-2 engine -- the second of which will boost the stage into an Earth-escape trajectory and into solar orbit.

Other first-day mission activities include a docked service propulsion engine burn to improve orbital lifetime and to test the command/service module (CSM) digital autopilot (DAD) during service propulsion system (SPS) burns.

The digital autopilot will undergo additional stability tests during the second work day when the SPS engine is ignited three more times. Also, the three docked burns reduce CSM weight to enhance possible contingency rescue of the lunar module during rendezvous using the service module reaction control thrusters.

A thorough checkout of lunar module systems takes up most of the third work day when the spacecraft commander and lunar module pilot transfer through the docking tunnel to the LM and power it up.

-more-

APOLLO 9

LAUNCH DAY

ORBITAL INSERTION

FLIGHT CREW PREPARATION

103 N. MILE ORBIT

DOCKING

DOCKED SPS BURN

SEPARATION

APOLLO 9   SECOND DAY

PITCH MANEUVER

HIGH APOGEE ORBITS

LANDMARK TRACKING

YAW-ROLL MANEUVER

LM SYSTEM EVALUATION

APOLLO 9 THIRD DAY

CREW TRANSFER

APOLLO 9 FOURTH DAY

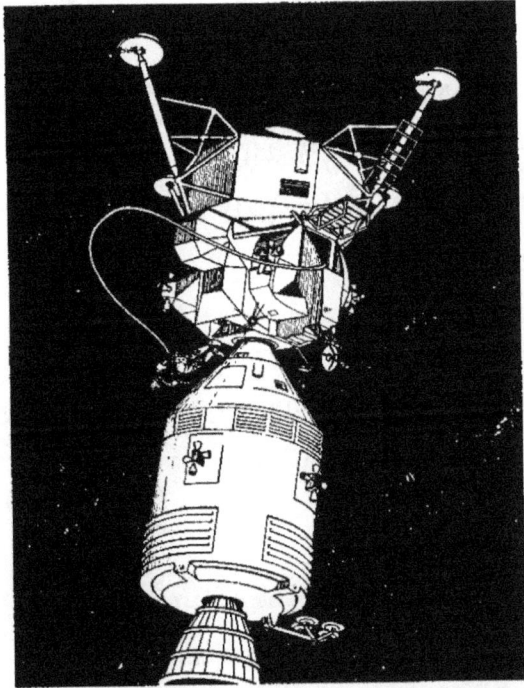

DAY-NIGHT EVA

TV - TEXAS, FLORIDA

CAMERA

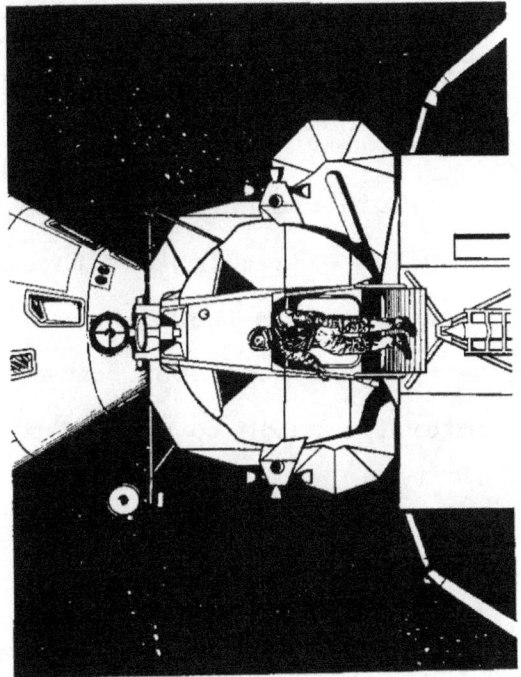

GOLDEN SLIPPERS

Among LM tests will be an out-of-plane docked burn of
the descent stage engine under control of the LM digital
autopilot, with the last portion of the burn manually
throttled by the spacecraft commander.

After LM power-down and crew transfer back to the command
module, the fifth docked service propulsion engine burn will
circularize the orbit at 133 nautical miles as a base orbit
for the LM active rendezvous two days later.

The fourth mission work day consists of further lunar
module checks and extravehicular activity. The spacecraft
commander and lunar module pilot again crawl through the tunnel
to power the LM and prepare the LM pilot's extravehicular mobility unit (EMU -- EVA suit with life support backpack) for
his two-hour stay outside the spacecraft.

Both spacecraft will be depressurized for the EVA and the
LM pilot will climb out through the LM front hatch onto the
"porch". From there, he will transfer hand-over-hand along
handrails to the open command module side hatch and partly
enter the cabin to demonstrate a contingency transfer capability.

The LM pilot will retrieve thermal samples on the spacecraft exterior and, returning to "golden slipper" foot restraints on the LM porch, will photograph both spacecraft from various angles and test the lunar surface television camera for about 10 minutes during a pass over the United States. This is a new model camera that has not been used in previous missions.

Both spacecraft will be closed and repressurized after the LM pilot gets into the LM, the LM will be powered down, and both crewmen will return to the command module.

The commander and LM pilot return to the LM the following day and begin preparations for undocking and a sequence simulating the checkout for a lunar landing descent.

A small thrust with the CSM reaction control system thrusters after separation from the LM places the CSM in an orbit for a small-scale rendezvous, "mini-football", in which the maximum distance between the two spacecraft is about three nautical miles. The LM rendezvous radar is locked on to the CSM transponder for an initial test during this period.

One-half revolution after separation, the LM descent engine is ignited to place the LM in an orbit ranging 11.8 nautical miles above and below the CSM orbit, and after $1\frac{1}{4}$ revolutions it is fired a second time, to place the LM in an orbit 11 nautical miles above the CSM.

The LM descent stage will be jettisoned and the LM ascent stage reaction control thrusters will next be fired to lower LM perigee to 10 miles below the CSM orbit and set up conditions for circularization. Maximum LM-to-CSM range will be 95 nautical miles during this sequence.

Circularization of the ascent stage orbit 10 nautical miles below the CSM and closing range will be the first duty in the mission for the 3,500-pound thrust ascent engine.

As the ascent stage approaches to some 20 nautical miles behind and 10 nautical miles below the CSM, the commander will thrust along the line of sight toward the CSM with the ascent stage RCS thrusters, making necessary midcourse corrections and braking maneuvers until the rendezvous is complete.

When the two vehicles dock, the ascent stage crew will prepare the ascent stage for a ground-commanded ascent engine burn to propellant depletion and transfer back into the command module. After the final undocking, the CSM will maneuver to a safe distance to observe the ascent engine depletion burn which will place the LM ascent stage in an orbit with an estimated apogee of 3,200 nautical miles.

The sixth mission day is at a more leisurely pace, with the major event being a burn of the service propulsion system to lower perigee to 95 nautical miles for improved RCS thruster deorbit capability.

APOLLO 9 FIFTH DAY

MAXIMUM SEPARATION

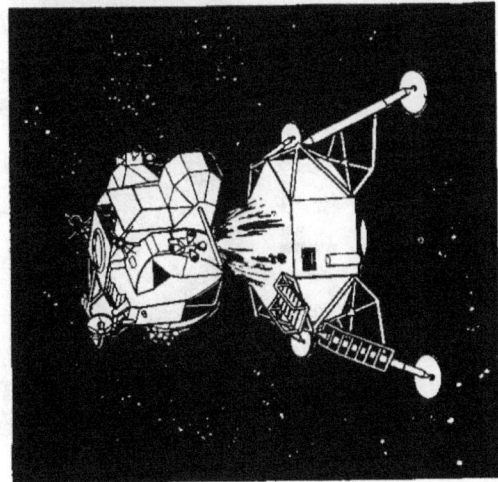

LM JETTISON ASCENT BURN

LM - BURNS FOR RENDEZVOUS

FORMATION FLYING AND DOCKING

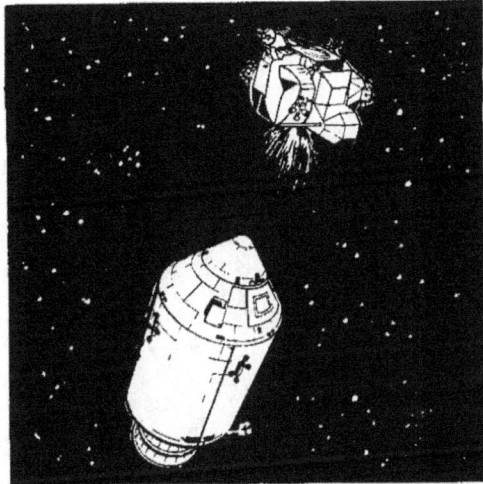

VEHICLES UNDOCKED

APS BURN

APOLLO 9   SIXTH THRU NINTH DAYS

LANDMARK SIGHTINGS, PHOTOGRAPH SPECIAL TESTS

SERVICE PROPULSION BURNS

The seventh SPS burn is scheduled for the eighth day to extend orbital lifetime and enhance RCS deorbit capability by raising apogee to 210 nautical miles.

The major activities planned during the sixth through tenth mission work days include landmark tracking exercises, spacecraft systems exercises, and a multispectral terrain photography experiment for Earth resources studies.

The eleventh work period begins with stowage of onboard equipment and preparations for the SPS deorbit burn 700 miles southeast of Hawaii near the end of the 150th revolution. Splashdown for a 10-day mission will be at 9:46 a.m. EST (238:46:30 GET) in the West Atlantic some 250 miles ESE of Bermuda and 1,290 miles east of Cape Kennedy (30.1 degrees north latitude by 59.9 degrees west longitude).

The Apollo 9 crew and spacecraft will be picked up by the landing platform-helicopter (LPH) USS Guadalcanal. The crew will be airlifted by helicopter the following morning to Norfolk, Va., and thence to the Manned Spacecraft Center, Houston. The spacecraft will be taken off the Guadalcanal at Norfolk, deactivated, and flown to the North American Rockwell Space Division plant in Downey, Calif., for postflight analysis.

The Saturn V launch vehicle with the Apollo spacecraft on top stands 363 feet tall. The five first stage engines of Saturn V develop a combined thrust of 7,720,174 pounds at first motion. Thrust increases with altitude until the total is 9,169,560 pounds an instant before center engine cutoff, scheduled for 2 minutes 14 seconds after liftoff. At that point, the vehicle is expected to be at an altitude of about 26 nm (30 sm, 45 km) and have a velocity of about 5,414 f/sec (1,650 m/sec, 3,205 knots, 3,691 mph). At first stage ignition, the space vehicle will weigh 6,486,915 pounds.

Apollo/Saturn V vehicles were launched Nov. 9, 1967, April 4, 1968, and Dec. 21, 1968, on Apollo missions. The last vehicle carried the Apollo 8 crew, the first two were unmanned.

The Apollo 9 Saturn V launch vehicle is different from the previous Saturn V's in the following aspects:

Dry weight of the first stage has been reduced from 304,000 to 295,600 pounds.

The first stage fueled weight at ignition has been increased from 4,800,000 to 4,946,337 pounds.

Instrumentation measurements in the first stage have been reduced from 891 to 648.

The camera instrumentation electrical power system was not installed on the first stage, and the stage carries neither a film nor television camera system.

The second stage will be somewhat lighter and slightly more powerful than previous S-II's. Maximum vacuum thrust for J-2 engines was increased from 225,000 to 230,000 pounds each. This changed second stage total thrust from 1,125,000 to 1,150,000 pounds. The maximum S-II thrust on this flight is expected to be 1,154,254 pounds.

The approximate dry weight of the S-II has been reduced from 88,000 to 84,600 pounds. The interstage weight was reduced from 11,800 to 11,664 pounds. Weight of the stage fueled has been increased from 1,035,000 to 1,069,114 pounds.

The S-II instrumentation system was changed from research and development to operational, and instrumentation measurements were reduced from 1,226 to 927.

Major differences between the S-IVB used on Apollo 8 and the one for Apollo 9 include:

Dry stage weight decreased from 26,421 to 25,300 pounds. This does not include the 8,081-pound interstage section. Fueled weight of the stage has been decreased from 263,204 to 259,337 pounds.

APOLLO 9

TENTH DAY

RE-ENTRY

CM/SM SEPARATION

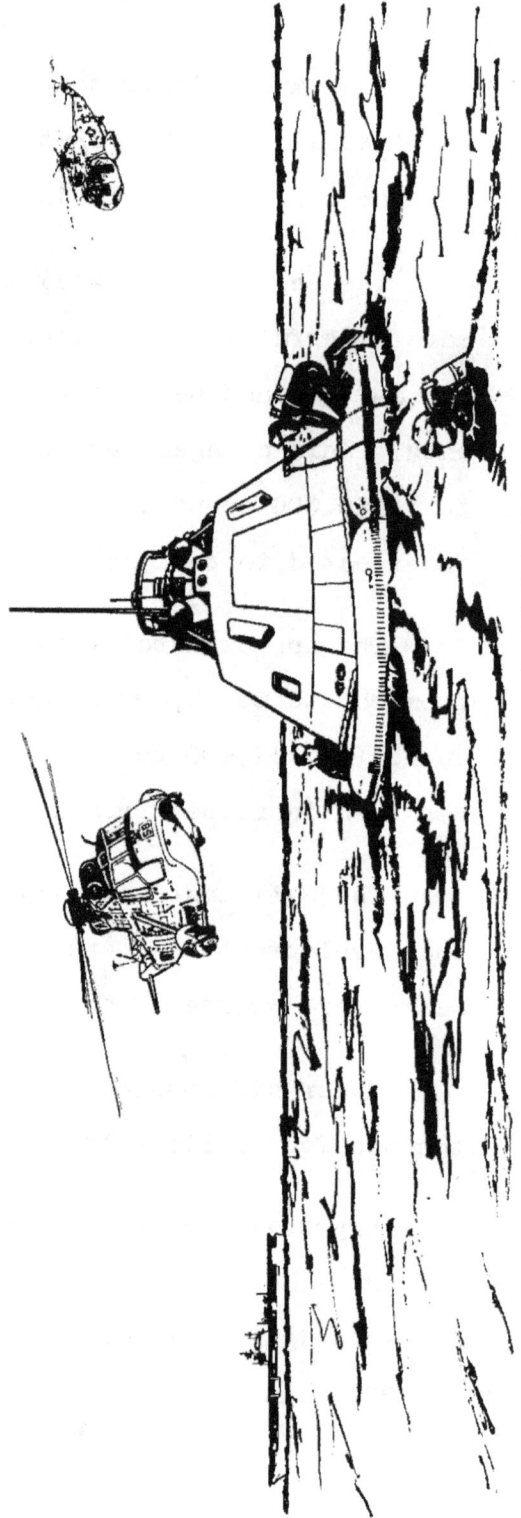

ATLANTIC - SPLASHDOWN

Stage measurements evolved from research and development to operational status, and instrumentation measurements were reduced from 342 to 280.

In the instrument unit, the rate gyro timer, thermal probe, a measuring distributor, tape recorder, two radio frequency assemblies, a source follower, a battery and six measuring racks have been deleted. Instrumentation measurements were reduced from 339 to 221.

During the Apollo 9 mission, communications between the spacecraft and the Mission Control Center, the spacecraft will be referred to as "Apollo 9" and the Mission Control Center as "Houston". This is the procedure followed in past manned Apollo missions.

However, during the periods when the lunar module is manned, either docked or undocked, a modified call system will be used.

Command Module Pilot David Scott in the command module will be identified as "Gumdrop" and Spacecraft Commander James McDivitt and Lunar Module Pilot Russell Schweickart will use the call sign "Spider."

Spider, of course is derived from the bug-like configuration of the lunar module. Gumdrop is derived from the appearance of the command and service modules when they are transported on Earth. During shipment they were wrapped in blue wrappings giving the appearance of a wrapped gumdrop.

(END OF GENERAL RELEASE; BACKGROUND INFORMATION FOLLOWS)

-more-

APOLLO 9 COUNTDOWN

The clock for the Apollo 9 countdown will start at
T-28 hours, with a six hour built-in-hold planned at T-9
hours, prior to launch vehicle propellant loading.

The countdown is preceded by a pre-count operation
that begins some 5½ days before launch. During this period
the tasks include mechanical buildup of both the command/
service module and LM, fuel cell activation and servicing and
loading of the super critical helium aboard the LM descent
stage. A 5½ hour built-in-hold is scheduled between the end
of the pre-count and start of the final countdown.

Following are some of the highlights of the final count:

T-28 hrs.	--Official countdown starts
T-27 hrs.	--Install launch vehicle flight batteries (to 23 hrs.) --LM stowage and cabin closeout (to 15 hrs.)
T-24 hrs. 30 mins.	--Launch vehicle systems checks (to 18:30 hrs.)
T-20 hrs.	--Top off LM super critical helium (to 17 hrs.)
T-16 hrs.	--Launch vehicle range safety checks (to 15 hrs.)
T-11 hrs. 30 mins.	--Install launch vehicle destruct devices --Command/service module pre-ingress operations
T-10 hrs. 30 mins.	--Start mobile service structure move to park site
T-9 hrs.	--Start six hour built-in-hold
T-9 hrs. counting	--Clear blast area for propellant loading
T-8 hrs. 15 mins.	--Launch vehicle propellant loading, three stages (liquid oxygen in first stage; liquid oxygen and liquid hydrogen in second, third stages). Continues thru T-3:30 hrs.

T-3 hrs. 10 mins.	--Spacecraft closeout crew on station
T-2 hrs. 40 mins.	--Start flight crew ingress
T-1 hr. 55 mins.	--Mission Control Center-Houston/ spacecraft command checks
T-1 hr. 50 mins.	--Abort advisory system checks
T-1 hr. 46 mins.	--Space vehicle Emergency Detection System (EDS) test
T-43 mins.	--Retrack Apollo access arm to stand-by position (12 degrees)
T-42 mins.	--Arm launch escape system
T-40 mins.	--Final launch vehicle range safety checks
T-30 mins.	--Launch vehicle power transfer test
T-20 mins.	--LM switch over to internal power
T-15 mins.	--Spacecraft to internal power
T-6 mins.	--Space vehicle final status checks
T-5 mins. 30 sec.	--Arm destruct system
T-5 mins.	--Apollo access arm fully retracted
T-3 mins. 10 sec.	--Initiate firing command (automatic sequencer)
T-50 sec.	--Launch vehicle transfer to internal power
T-8.9 sec.	--Ignition sequence start
T-2 sec.	--All engines running
T-0	--Liftoff

*NOTE: Some changes in the above countdown are possible as a result of experience gained in the Countdown Demonstration Test (CDDT) which occurs about 10 days before launch.

SEQUENCE OF EVENTS

NOMINAL MISSION

Time from Liftoff (Hr:Min:Sec)	Date	Time (EST)		Event
00:00:00	Feb. 28	11:00	AM	First Motion
00:00:12				Tilt Initiation
00:01:21				Maximum Dynamic Pressure
00:02:14				S-IC Center Engine Cutoff
00:02:39		11:02:39	AM	S-IC Outboard Engine Cutoff
00:02:40				S-IC/S-II Separation
00:02:42		11:02:42	AM	S-II Ignition
00:03:10				S-II Aft Interstage Separation
00:03:15		11:03:15	AM	Launch Escape Tower Jettison
00:03:21				Initiate IGM
00:08:53		11:08:53	AM	S-II Cutoff
00:08:54				S-II/S-IVB Separation
00:08:57		11:08:57	AM	S-IVB Ignition
00:10:49		11:10:49	AM	S-IVB First Cutoff
00:10:59		11:10:59	AM	Insertion into Earth Parking Orbit
02:34:00		1:34	PM	S-IVB Enters Transposition Attitude
02:43:00		1:43	PM	Spacecraft Separation
02:53:43		1:53	PM	Spacecraft Docking
04:08:57		3:08	PM	Spacecraft Final Separation
04:11:25		3:11	PM	CSM RCS Separation Burn

Time from Liftoff	Date	Time		Event
04:45:41		3:45	PM	S-IVB Reignition
04:46:43		3:46	PM	S-IVB Second Cutoff
04:46:53				S-IVB Insertion into Intermediate Orbit
06:01:40		5:01	PM	SPS Burn No. 1 (Docked)
06:07:04		5:07	PM	S-IVB Reignition
06:11:05		5:11	PM	S-IVB Third Cutoff
06:11:15				S-IVB Insertion into Escape Orbit
06:12:36		5:12	PM	Start S-IVB LOX Dump
06:23:46		5:23	PM	S-IVB LOX Dump Cutoff
06:23:56				Start S-IVB LH2 Dump
06:42:11		5:42	PM	S-IVB LH2 Dump Cutoff
22:12:00	Mar. 1	9:12	AM	SPS Burn No. 2 (Docked)
25:18:30		12:18	PM	SPS Burn No. 3 (Docked)
28:28:00		3:28	PM	SPS Burn No. 4 (Docked)
40:00:00	Mar. 2	3:00	AM	LM Systems Evaluation
46:29:00		(:29	PM	LM TV Transmission
49:43:00		12:43	PM	Docked LM Descent Engine Burn
54:26:16		5:26	PM	SPS Burn No. 5 (Docked)
68:00:00	Mar. 3	7:00	AM	Begin Preps for EVA
73:10:00		12:10	PM	EVA by LM Pilot
75:05:00		2:05	PM	EVA TV
75:20:00		2:20	PM	EVA Ends
89:00:00	Mar. 4	4:00	AM	Rendezvous Preps Begin

Time from Liftoff	Date	Time		Event
93:05:45	Mar. 4	8:05	AM	CSM RCS Separation Burn
93:50:03		8:50	AM	LM Descent Engine Phasing Burn
95:41:48		10:41	AM	LM Descent Engine Insertion Burn
96:22:00		11:22	AM	LM RCS Concentric Sequence Burn
97:06:28		12:06	PM	LM Ascent Engine Circularization Burn
97:59:21		12:59	PM	LM RCS Terminal Phase Burn
98:31:41		1:31	PM	Terminal Phase Finalization Docking
100:26:00		3:26	PM	LM APS Long Duration Burn
121:59:00	Mar. 5	12:59	PM	SPS Burn No. 6 (CSM only)
169:47:00	Mar. 7	12:47	PM	SPS Burn No. 7
238:10:47	Mar. 10	9:10	AM	SPS Burn No. 8 (Deorbit)
238:41:40		9:41	AM	Main Parachute Deployment
238:46:30		9:46	AM	Splashdown

APOLLO 9 (AS-504) MISSION PROFILE

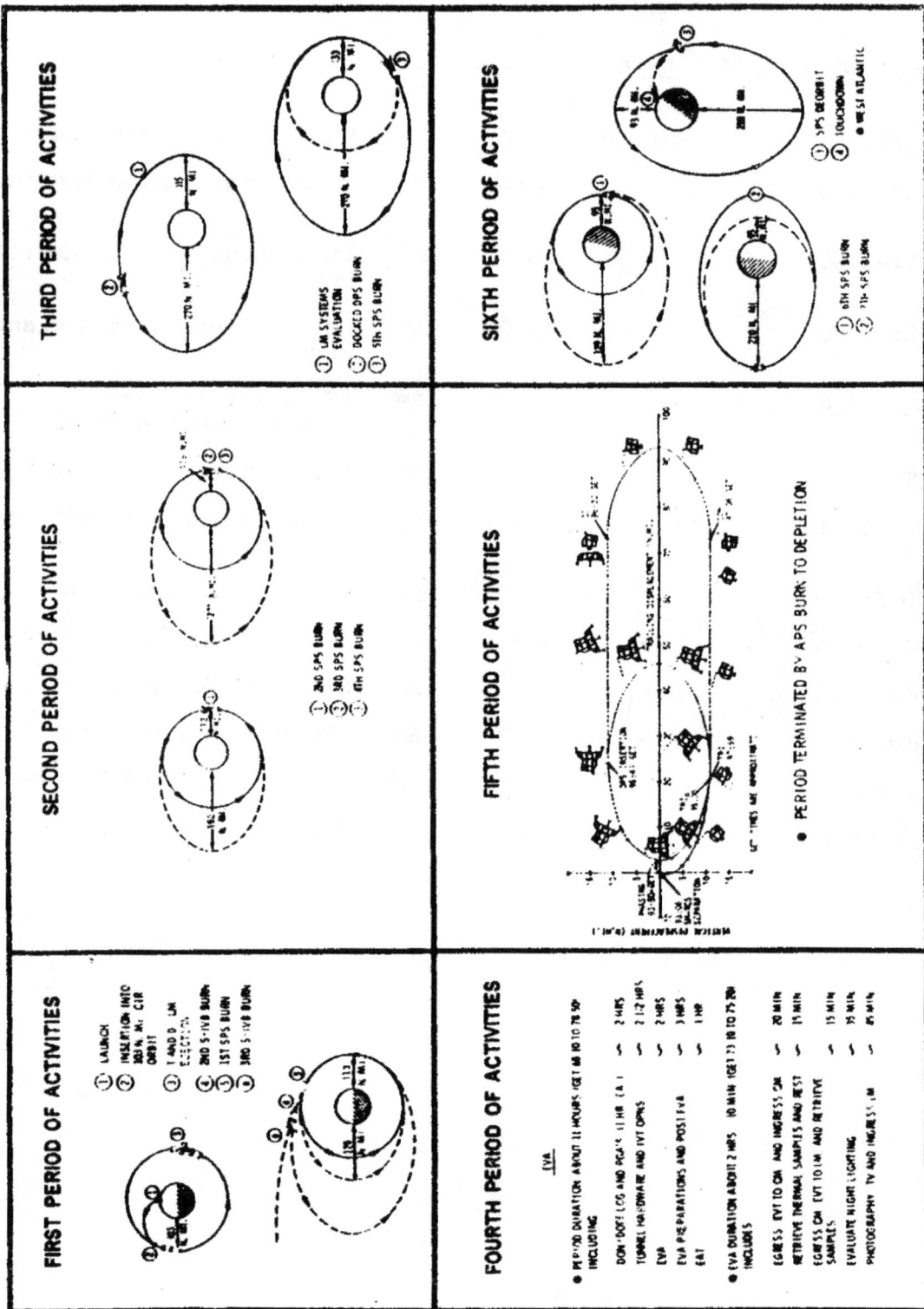

APOLLO 9 MISSION OBJECTIVES

The Apollo 8 lunar orbit mission in December fully demonstrated that the Apollo command and service modules are capable of operating at lunar distances. But a missing link in flying an Apollo spacecraft to the Moon for a lunar landing is a manned flight in the lunar module.

Apollo 9's primary objective will be to forge that missing link with a thorough checkout in Earth orbit of the lunar module and its systems in a series of tests including maneuvers in which the LM is the active rendezvous vehicle -- paralleling an actual lunar orbit rendezvous.

Apollo 9 will be the most ambitious manned space flight to date, including the Apollo 8 lunar orbit mission. Many more tests are packed into the mission, and most of these deal with the yet-untried lunar module. While the lunar module has been flown unmanned in space (Apollo 5/LM-1, Jan. 22, 1968), the real test of a new spacecraft type comes when it is flown manned. Many of the planned tests will exceed the conditions that will exist in a lunar landing mission.

Although Apollo 9 will be followed by a lunar orbit mission in which the LM descends to 50,000 feet above the Moon's surface, but does not land, there will not be any other long-duration burns of the descent engine before the first lunar landing -- possibly on Apollo 11.

Top among mission priorities are rendezvous and docking of the command module and the lunar module. The vehicles will dock twice -- once when the LM is still attached to the S-IVB, and again following the rendezvous maneuver sequence. The dynamics of docking the vehicles can be likened to coupling two freight cars in a railroad switching yard -- but using a coupling mechanism built with the precision of a fine watch.

Next in mission priority are special tests of lunar module systems, such as performance of the descent and ascent engines in various guidance control modes and the LM environmental control and electrical power systems that can only get final checkout in space.

The preparations aboard the LM for EVA -- checkout of the LM pilot's extravehicular mobility unit and configuring the LM to support EVA are also of high priority. The Apollo 9 EVA will be the only planned EVA in the Apollo program until the first lunar landing crewmen climb down the LM front leg to walk upon the lunar surface.

MISSION TRAJECTORY AND MANEUVER DESCRIPTION

(Note: Information presented herein is based upon
a nominal mission and is subject to change prior
mission or in real time during the mission to meet
changing conditions.)

Launch

Apollo 9 is scheduled to be launched at 11 a.m. EST from
NASA Kennedy Space Center Launch Complex 39A on a 72-degree
azimuth and inserted into a 103 nm (119 sm, 191.3 km) circular
Earth orbit by Saturn V launch vehicle No. 504. Insertion will
take place at 10 minutes 59 seconds after liftoff.

Transposition and Docking

Following insertion into orbit, the S-IVB third stage
maintains an attitude level with the local horizontal while
the Apollo 9 crew conducts post-insertion CSM systems checks
and prepares for a simulated S-IVB translunar injection restart.

At 2 hours 34 minutes ground elapsed time (GET) the S-IVB
enters transposition and docking attitude (15 degree pitch, 35
degree yaw south); and the CSM separates from the S-IVB at 2
hours 43 minutes GET at one fps to about 50 feet separation,
where velocity is nulled and the CSM pitches 180 degrees and
closes to near the lunar module docking collar for station
keeping. Docking is completed at about 2 hours 53 minutes GET
and the LM is pressurized with the command module surge tanks
and reentry bottles.

LM Ejection and Separation

The lunar module is ejected from the spacecraft/LM adapter
by spring devices at the four LM landing gear "knee" attach
points. A three-second SM RCS burn separates the CSM/LM for
crew observation of the first S-IVB restart.

S-IVB Restarts

After spacecraft separation, the S-IVB resumes a local
horizontal attitude for the first J-2 engine restart at 4:45:41
GET. The CSM/LM maintains a separation of about 2,000 feet
from the S-IVB for the restart. A second S-IVB restart at
6:07:04 GET followed by propellant dumps place the S-IVB in an
Earth escape trajectory and into solar orbit.

CMS/S-IVB ORBITAL OPERATIONS

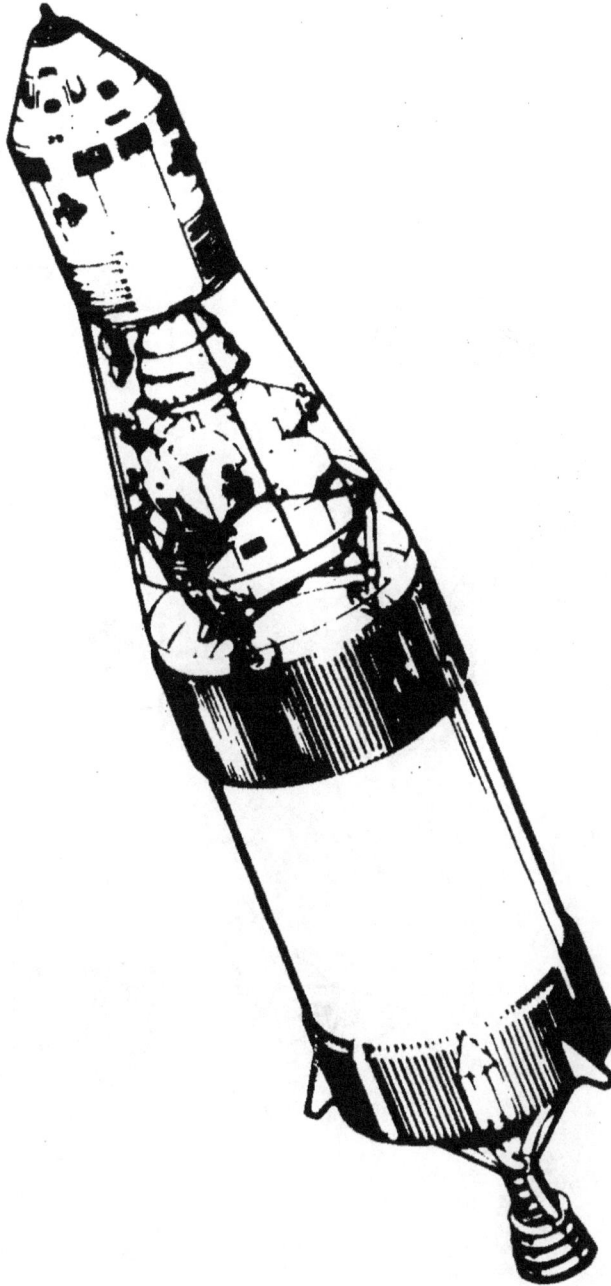

NASA HQ MA69-4177
1-28-69

TRANSPOSITION, DOCKING AND CSM/LM EJECTION

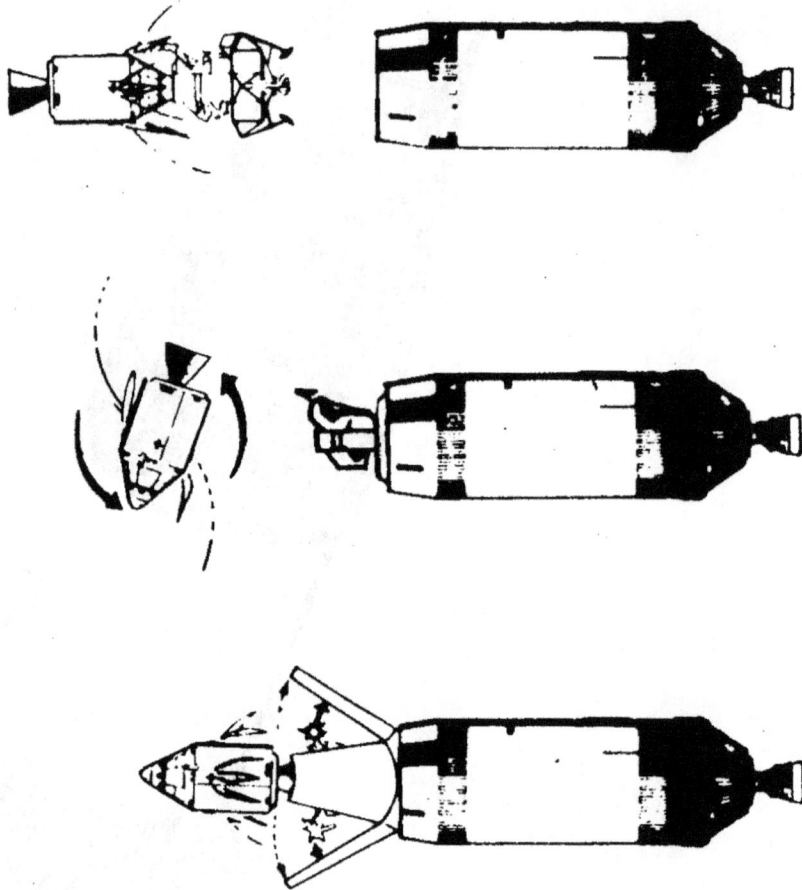

NASA HQ MA69-4178
1-28-69

S-IVB SOLO OPERATIONS

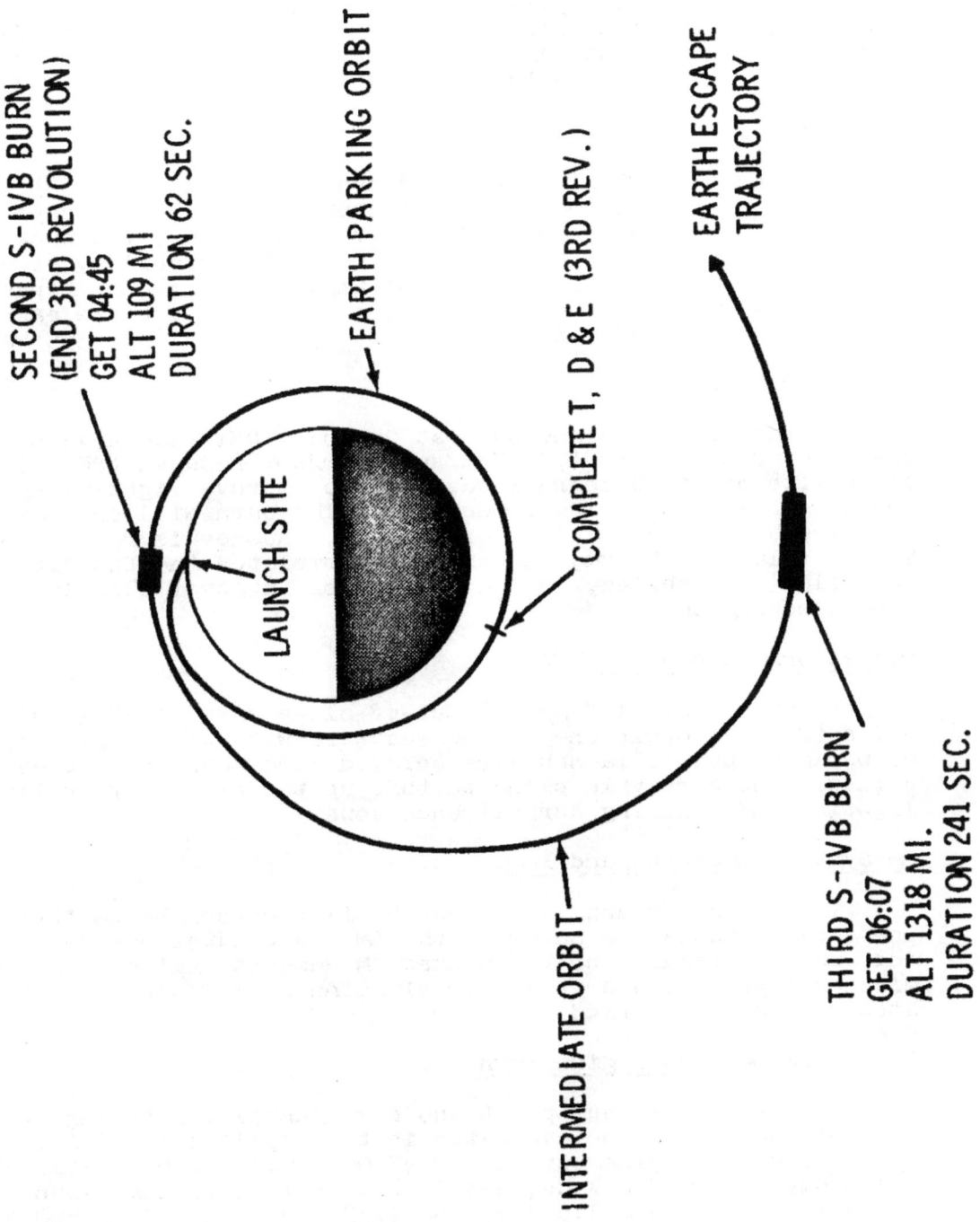

SECOND S-IVB BURN
(END 3RD REVOLUTION)
GET 04:45
ALT 109 MI
DURATION 62 SEC.

EARTH PARKING ORBIT

LAUNCH SITE

COMPLETE T, D & E (3RD REV.)

EARTH ESCAPE
TRAJECTORY

THIRD S-IVB BURN
GET 06:07
ALT 1318 MI.
DURATION 241 SEC.

INTERMEDIATE ORBIT

Service Propulsion System (SPS) Burn No. 1

A 36.8 fps (11.2 m/sec) docked SPS burn at 6:01:40 GET enhances spacecraft orbital lifetime and demonstrates stability of the CSM digital autopilot. The new orbit is 113 x 131 nm (130 x 151 sm, 290 x 243 km).

Docked SPS Burn No. 2

This burn at 22:12:00 GET reduces CSM weight by 7,355 pounds (3,339 kg) so that reaction control propellant usage in a contingency LM rescue (CSM-active rendezvous) would be lessened; provides continuous SM RCS deorbit capability; tests CSM digital autopilot in 40 percent amplitude stroking range. The burn is mostly out of plane adjusting orbital plane eastward (nominally) and raises apogee to 192 nm (221 sm, 355 km).

Docked SPS Burn No. 3

Another out-of-plane burn at 25:18:30 GET with a velocity change of 2,548.2 fps (776.7 m/sec) further reduces CSM weight and shifts orbit 10 degrees eastward to improve lighting and tracking coverage during rendezvous. The burn will consume 18,637 pounds (8,462 kg) of propellant. Apogee is raised to 270 nm (310 sm, 500 km) and the burn completes the CSM digital autopilot test series. A manual control takeover also is scheduled for this.

Docked SPS Burn No. 4

A 299.8 fps (91.4 m/sec) out-of-plane burn at 28:28:00 GET shifts the orbit one degree eastward without changing apogee or perigee, but if launch were delayed more than 15 minutes, the burn will be partly in-plane to tune up the orbit for optimum lighting and tracking during rendezvous.

LM Systems Checkout and Power-Up

The commander and lunar module pilot enter the LM through the docking tunnel to power-up the LM and conduct systems checkout, and to prepare for the docked LM descent engine burn. Premission intravehicular transfer timeliness will also be evaluated during this period.

Docked LM Descent Engine Burn

The LM digital autopilot and crew manual throttling of the descent engine will be evaluated in this 1,714.1 fps (522.6 m/sec) descent engine burn at 49:43:00 GET. The burn will shift the orbit eastward 6.7 degrees (for a nominal on-time launch) and will result in a 115 x 270 nm (132 x 310 sm, 213 x 500 km) orbit. After the burn, the LM crew will power down the LM and return to the command module.

-more-

NASA HQ MA69-4179
1-28-69

DOCKED SPS BURNS

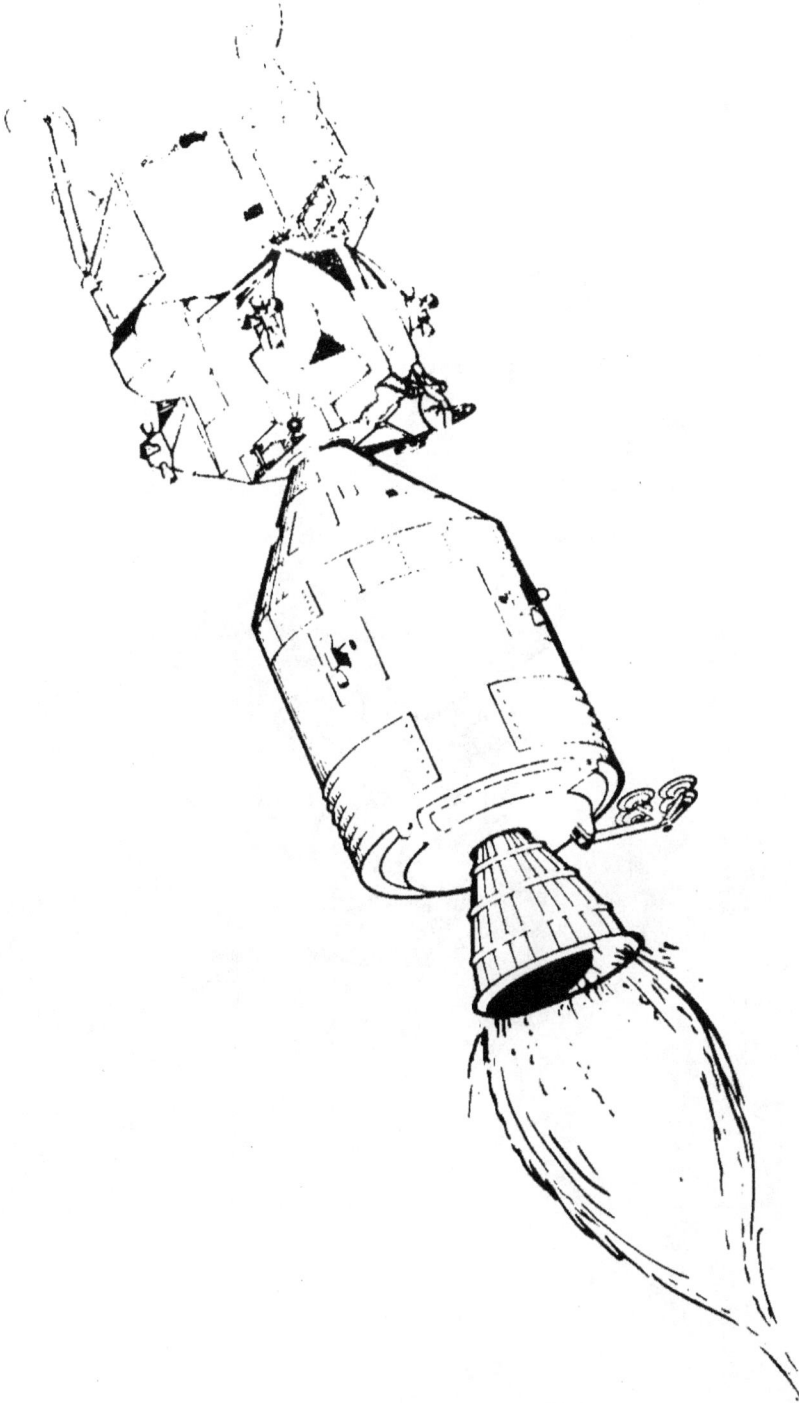

NASA HQ MA68-4180
1-2-__

DOCKED DPS BURN

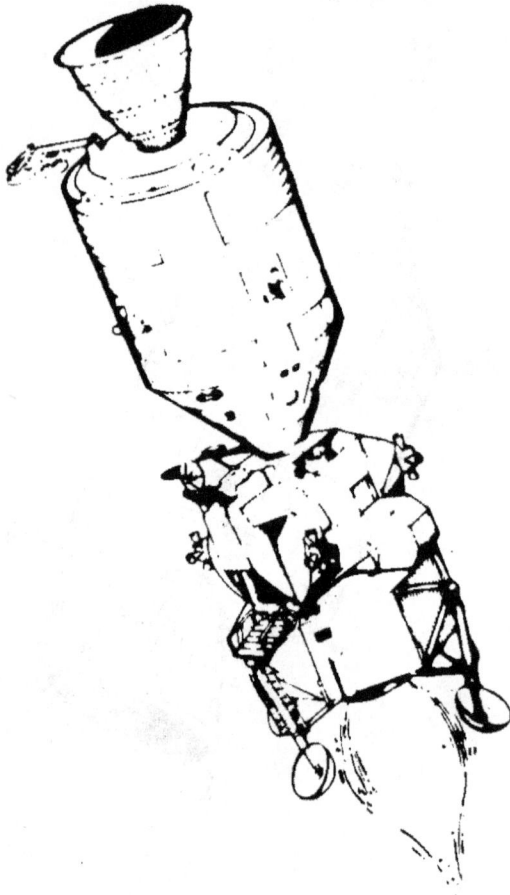

Docked SPS Burn No. 5

Apollo 9's orbit is circularized to 133 nm (153 sm, 246 km) by a 550.6 fps (167.8 m/sec) SPS burn at 54:26:16 GET. The 133 nm circular orbit becomes the base orbit for the LM-active rendezvous.

Extravehicular Activity

The commander and the LM pilot will transfer to the LM through the docking tunnel at about 68 hours GET and power up the spacecraft and prepare for EVA. The LM pilot will don the extravehicular mobility unit (EMU), check it out, and leave the LM through the front hatch at 73:10:00 GET. Tethered by a nylon rope, the LM pilot will move with the aid of handrails to the open command module side hatch and place his lower torso into the cabin to demonstrate EVA LM crew rescue. He then will return to the LM "porch" collecting thermal samples from the spacecraft exterior enroute. The LM pilot, restrained by "golden slippers" on the LM porch, photographs various components of the two spacecraft. The commander will pass the LM television camera to the LM pilot who will operate it for about 10 minutes beginning at 75:20:00 GET during a stateside pass. The LM pilot will enter the LM at 75:25:00 GET through the front hatch and the spacecraft will be repressurized. Both crewmen will transfer to the command module after powering down the LM systems.

CSM RCS Separation Burn

The commander and the LM pilot transfer to the LM at about 89 hours GET and begin LM power-up and preparations for separation and LM-active rendezvous. The first in the series of rendezvous maneuvers is a 5 fps (1.5 m/sec) radially downward CSM RCS burn at 93:05:45 GET which places the CSM in a 131 x 132 nm (151 x 152 sm, 243 x 245 km) equiperiod orbit for the "mini-football" rendezvous with a maximum vehicle separation of less than two nautical miles. During the mini-football, the LM radar is checked out and the LM inertial measurement unit (IMU) is aligned.

LM Descent Engine Phasing Burn

An 85 fps (25.9 m/sec) LM descent engine burn radially upward at 93:50:03 GET places the LM in an equiperiod 119 x 145 nm (137 x 167 sm, 220 x 268 km) orbit with apogee and perigee about 11.8 miles above and below, respectively, those of the command module. Maximum range following this maneuver is 48 nm and permits an early terminal phase maneuver and rendezvous if the full rendezvous sequence for some reason is no-go. The burn is the only one that is controlled by the LM abort guidance system with the primary system acting as a backup.

NASA HQ MA69-4181
1-28-69

EXTRAVEHICULAR ACTIVITY

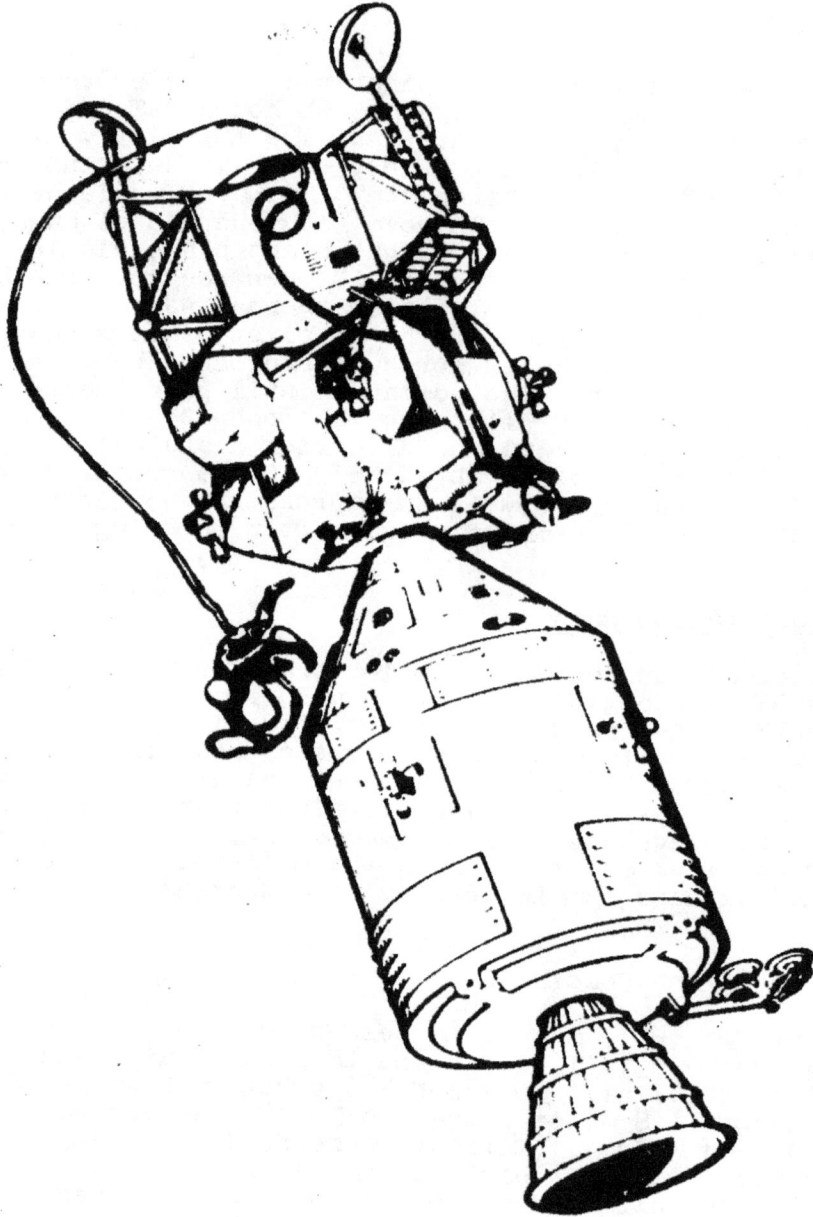

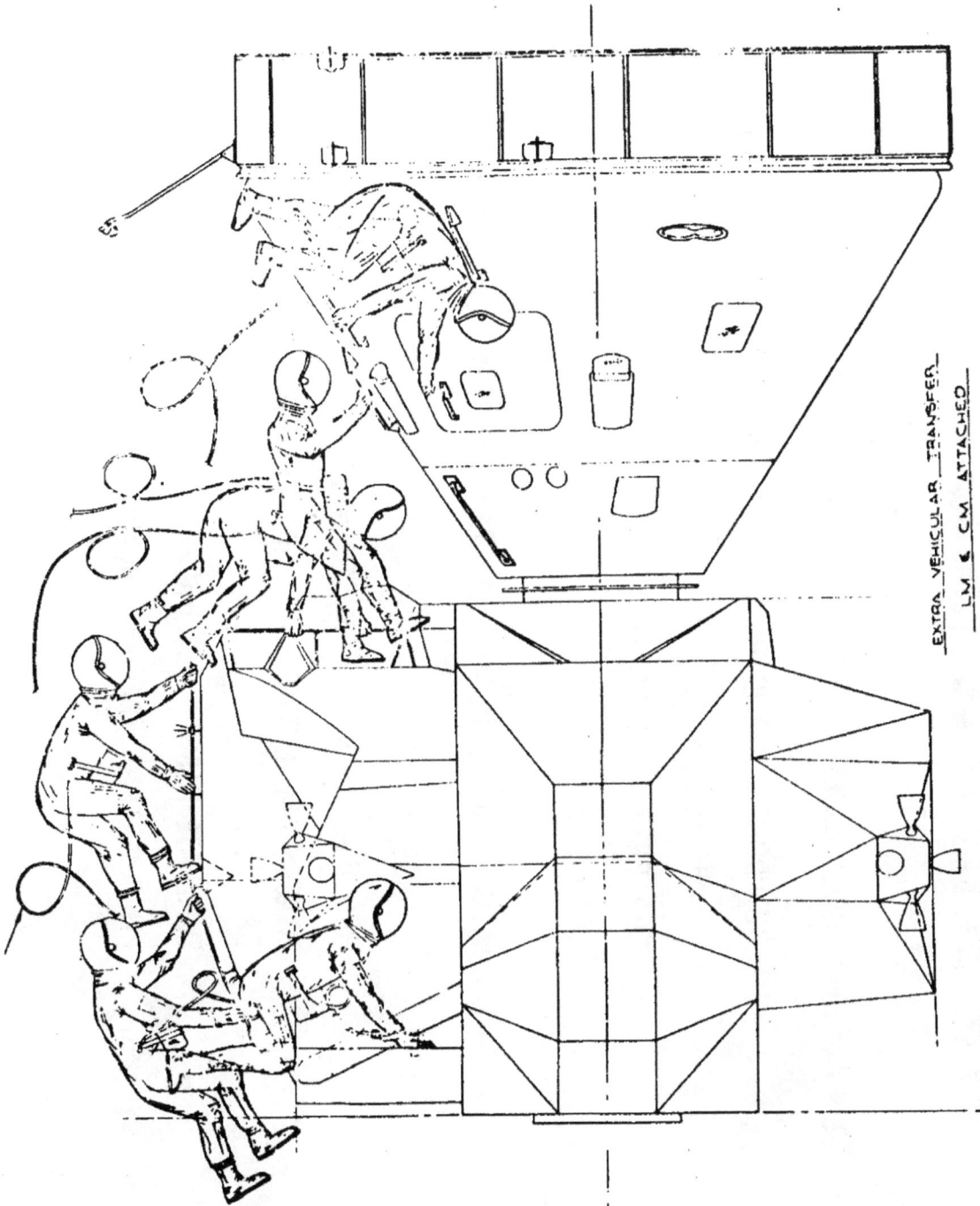

EXTRA VEHICULAR TRANSFER
LM & CM ATTACHED

EVA SCHEDULE (MISSION DAY 4)

- ABOUT 11 HOURS (GET 68:10 TO 78:50) INCLUDING:

DON/DOFF LCG AND PGA'S (1 HR. EA.)	—	2 HRS
TUNNEL HARDWARE AND IVT OPNS	—	2 1/2 HRS
EVA PREPARATIONS AND POST EVA	—	3 HRS
EAT	—	1 HR

- EVA DURATION 2 HRS., 10 MIN (GET 73:10 TO 75:20) INCLUDES:

EGRESS, EVT TO CM, AND INGRESS CM	—	20 MIN.
RETRIEVE THERMAL SAMPLES AND REST	—	15 MIN.
EGRESS CM, EVT TO LM, AND RETRIEVE SAMPLES	—	15 MIN.
EVALUATE NIGHT LIGHTING	—	35 MIN.
PHOTOGRAPHY, TV, AND INGRESS LM	—	45 MIN.

RENDEZVOUS MANEUVERS AND BACKUP

MANEUVER	SEPARATION	PHASING	INSERTION	CSI	CDH	TPI
GET METHOD DIRECTION DURATION VELOCITY	93:05 CSM/RCS CSM-DOWN 10.9 SEC. 5 FPS	93:50 DPS-AGS UP 25.2 SEC. 85 FPS	95:42 DPS-PGNCS POSIGRADE 24.8 SEC. 40 FPS	96:22 LM/RCS RETROGRADE 30.6 SEC. 38 FPS	97:06 APS-PGNCS RETROGRADE 3.1 SEC. 38 FPS	97:59 LM/RCS POSIGRADE 17.6 SEC. 22 FPS
BACKUP BURNS	NONE REQUIRED		CSM INSERTION - CANCELLING BURN	CSM MIRROR IMAGE BURN	LM-RCS OR CSM MIRROR IMAGE BURN	CSM MIRROR IMAGE BURN
ABORT CAPABILITY	CSM OR LM EXECUTE LINE - OF-SIGHT RANGE AND RANGE RATE CONTROL	CSM OR LM EXECUTE TPI OR LINE-OF- SIGHT RANGE RATE CON- TROL	TWO IMPULSE TECHNIQUE TO REDUCE RETURN TIME TO ABOUT 56 MIN.			DOCKING NOMINALLY AT 99:17

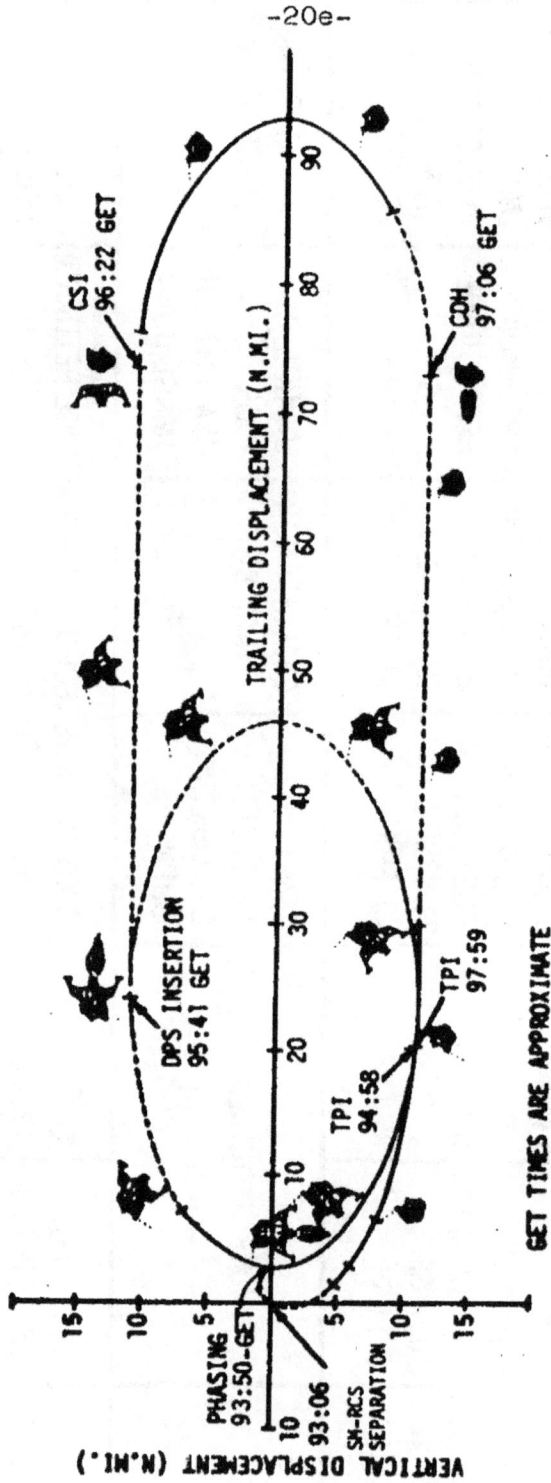

"D" MISSION RENDEZVOUS PROFILE
(CSM-CENTERED RELATIVE MOTION)

LM Descent Engine Insertion Burn

This burn at 95:41:48 GET of 39.9 fps (9.2 m/sec) is at a 10-percent throttle setting and places the LM in a 142 x 144 nm (163 x 166 sm, 263 x 265 km) near circular orbit and 11 miles above that of the CSM.

CSM Backup Maneuvers

For critical rendezvous maneuvers after the insertion burn, the CM pilot will be prepared to make a "mirror-image" burn of equal velocity but opposite in direction one minute after the scheduled LM maneuver time if for some reason the LM cannot make the maneuver. Such a CSM burn will cause the rendezvous to be completed in the same manner as if the LM had maneuvered nominally.

LM RCS Concentric Sequence Burn (CSI)

A retrograde 37.8 fps (815 m/sec) RCS burn at 96:22:00 GET lowers LM perigee to about 10 miles below that of the CSM. The LM RCS is interconnected to the ascent engine propellant tankage for the burn, and the descent stage is jettisoned prior to the start of the burn.

The CSI burn is nominally computed onboard the LM to cause the phase angle at constant delta height (CDH) to result (after CDH is performed) in proper time for terminal phase initiate (TPI).

LM Ascent Engine Circularization Burn (CDH)

The LM ascent stage orbit is circularized by a 37.9 fps (8.5 m/sec) retrograde burn at 97:06:28 GET at LM perigee. At the end of the burn, the LM ascent stage is about 75 miles behind the CSM and in a coelliptic orbit about 10 miles below that of the CSM. The new LM ascent stage orbit becomes 119 x 121 nm (137 x 139, 220 x 224 km).

LM RCS Terminal Phase Initialization Burn

At 97:59:21 GET, when the LM ascent stage is about 20 nm behind and 10 nm below the CSM, a LM RCS 21.9 fps (6.6 m/sec) burn along the line of sight toward the CSM begins the final rendezvous sequence. Midcourse corrections and braking maneuvers complete the rendezvous and docking is estimated to take place at 98:31:41.

LM ACTIVE RENDEZVOUS

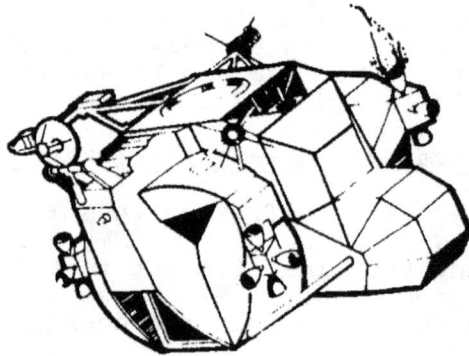

NASA HQ MA69-4182
1-28-69

APS BURN TO DEPLETION

NASA HQ MA0?-41ᵒᵒ
1-2ᵒ-0:

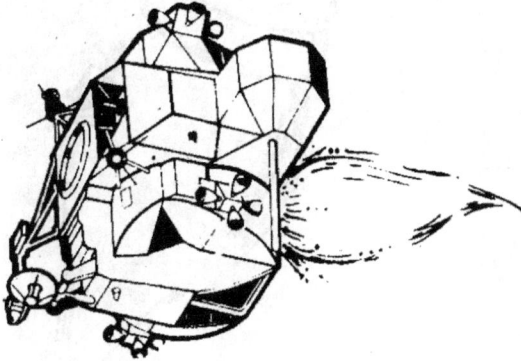

NASA HQ MA67-4164
1-26-67

CSM SOLO SPS BURNS

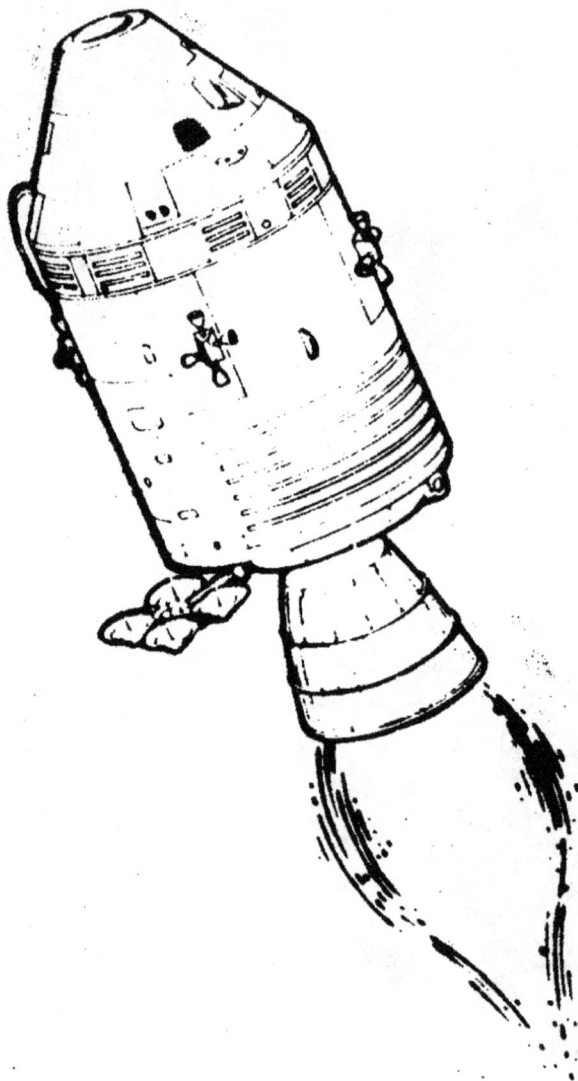

LM Ascent Engine Long-Duration Burn

Following docking and crew transfer back into the CSM, the LM ascent stage is jettisoned and the CSM maneuvers out of plane for separation. The LM ascent engine is ground commanded to burn to depletion at 100:26:00 GET, with LM RCS propellant augmented by the ascent tankage crossfeed. The burn will raise LM ascent stage apogee to about 3,200 nm (3,680 sm, 5,930 km). The burn is 5,658.5 fps (1,724.7 m/sec).

SPS Burn No. 6

At 121:59:00 GET, the SPS engine is ignited in a 66.1 fps (20.1 m/sec) retrograde burn lower perigee to 95 nm (109 sm, 176 km) to improve the spacecraft's backup capability to deorbit with the RCS thrusters.

SPS Burn No. 7

Apollo 9 orbital lifetime is extended and RCS deorbit capability improved by this 173.6 fps (52.6 m/sec)posigrade burn at 169:47:00 GET. The burn raises apogee to 210 nm (241 sm, 388 km). The burn also shifts apogee to the southern hemisphere to allow longer free-fall time to entry after a nominal SPS deorbit burn.

SPS Burn No. 8

The SPS deorbit burn is scheduled for 238:10:47 GET and is a 252.9 fps (77 m/sec) retrograde burn beginning about 700 nm southeast of Hawaii.

Entry

After deorbit burn cutoff, the CSM will be yawed 45 degrees out of plane for service module separation. Entry (400,000 feet) will begin about 15 minutes after deorbit burn, and main parachutes will deploy at 238:41:40 GET. Splashdown will be at 238:46:30 GET at 30.1 degrees north latitude, 59.9 degrees west longitude.

GEODETIC LATITUDE (DEG)

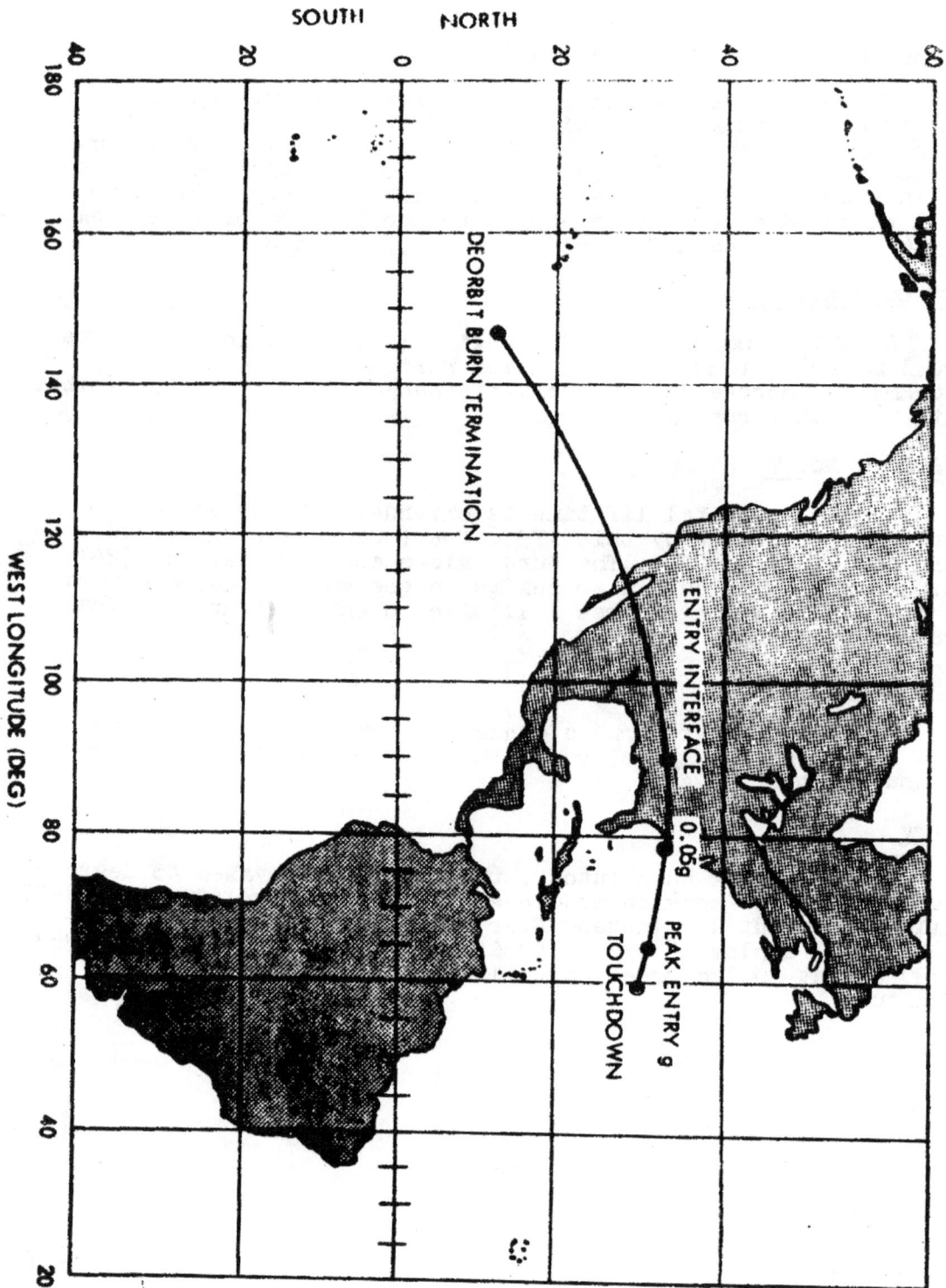

SOUTH   NORTH

WEST LONGITUDE (DEG)

DEORBIT BURN TERMINATION

ENTRY INTERFACE

0.05g

PEAK ENTRY g

TOUCHDOWN

APOLLO 9 MISSION EVENTS

Event	Ground Elapsed Time hrs:min:sec	Velocity Change fps (m/sec)	Purpose (Resultant Orbit)
Insertion	00:10:59	25,245.6 (7,389.7)*	Insertion into 103 nm (191.3 km) circular Earth parking orbit
CSM separation, docking	02:43:00	1 (.3)	Hard-mating of CSM and LM docking tunnel
LM ejection	04:08:57	.4 (.12)	Separates LM from S-IVB/SLA
CSM RCS separation burn	04:11:25	3 (.1)	Provides separation prior to S-IVB restart
Docked SPS burn No. 1	06:01:40	36.8 (11.2)	Orbital lifetime, first test of CSM digital autopilot (orbit: 113x128 nm)
Docked SPS burn No. 2	22:12:00	849.6 (259)	Reduces CSM weight for LM rescue and RCS deorbit; Second digital autopilot stroking test (113x192 nm)
Docked SPS burn No. 3	25:18:30	2,548.2 (776.7)	Reduces CSM weight; completes test of digital autopilot stroking (115x270 nm)
Docked SPS burn No. 4	28:28:00	299.8 (91.4)	Out-of-plane burn adjusts orbit for launch delays; no apogee/perigee change if on-time launch. Late launch: apogee between 130 and 500 nm.
LM Systems evaluation	40:00:00	---	Demonstrate crew intravehicular transfer to LM; power up and checkout LM
Docked LM descent engine burn	49:43:00	1,714.1 (522.6)	Demonstrate LM digital autopilot control, manual throttling of descent engine
Docked SPS burn No. 5	54:26:16	550.6 (167.8)	Circularizes orbit to 133 nm for rendezvous
Extravehicular activity	73:10:00	---	LM pilot demonstrates EVA transfer from LM to CSM and return; evaluates CM hatch, collects samples on spacecraft exterior, evaluates EVA lighting, performs EVA photography and TV test

-24-

Event	Ground Elapsed Time hrs:min:sec	Velocity fps(m/sec)	Purpose (Resultant Orbit)
CSM RCS separation burn	93:05:45	5.0 (1.5)	Radially downward burn to place CSM in equi-period orbit for mini-football rendezvous (131x132)
LM descent engine phasing burn	93:50:03	85 (25.9)	Demonstrates LM abort guidance system control of descent engine burn, provides radial separation for LM-active rendezvous (LM 119x145)
LM descent engine insertion burn	95:41:48	39.9 (9.2)	Adjusts height differential and thus phase angle between LM and CSM at time of CSI maneuver (LM 142x144)
LM RCS concentric sequence burn	96:22:00	37.8 (8.5)	Retrograde burn adjusts LM perigee to about 10 nm below that of CSM; LM is staged before burn
LM APS circularization burn	97:06:28	37.9 (8.5)	Retrograde burn circularizes LM orbit to about 10 nm below and 75 nm behind and closing on the CSM (LM 119x121)
LM RCS terminal phase burn	97:59:21	21.9 (6.6)	LM thrusts along line of sight toward CSM to begin intercept trajectory (LM 121x133)
Terminal phase finalization, rendezvous	98:31:41	28.4 (8.6)	Station keeping, breaking maneuvers to complete rendezvous (LM/CSM 131x133); docking should occur at 99:18
LM APS long duration burn (unmanned)	100:26:00	5,658.5 (1,724.7)	LM ascent engine ground commanded to burn to depletion after CSM undocks; demonstrates ascent engine ability for lunar landing mission profile (LM 131x3,200)
SPS burn No. 6 (CSM alone)	121:59:00	66.1 (20.1)	Lowers CSM perigee to enhance RCS deorbit capability (95x130)

Event	Ground Elapsed Time hrs:min:sec	Velocity fps (m/sec)	Purpose (Resultant Orbit)
SPS burn No. 7 (CSM alone)	169:47:00	173.6 (52.6)	Extends orbital lifetime and lessens RCS deorbit requirements by raising apogee (97x210)
*SPS burn No. 8 (CSM alone)	238:10:47	252.9 (77)	Deorbits spacecraft (29x210)
Main parachutes deploy	238:41:41	---	---
Splashdown	238:46:30	---	Landing at 30.1 degrees North Latitude x 59.9 degrees West Longitude

● This figure is orbital insertion velocity

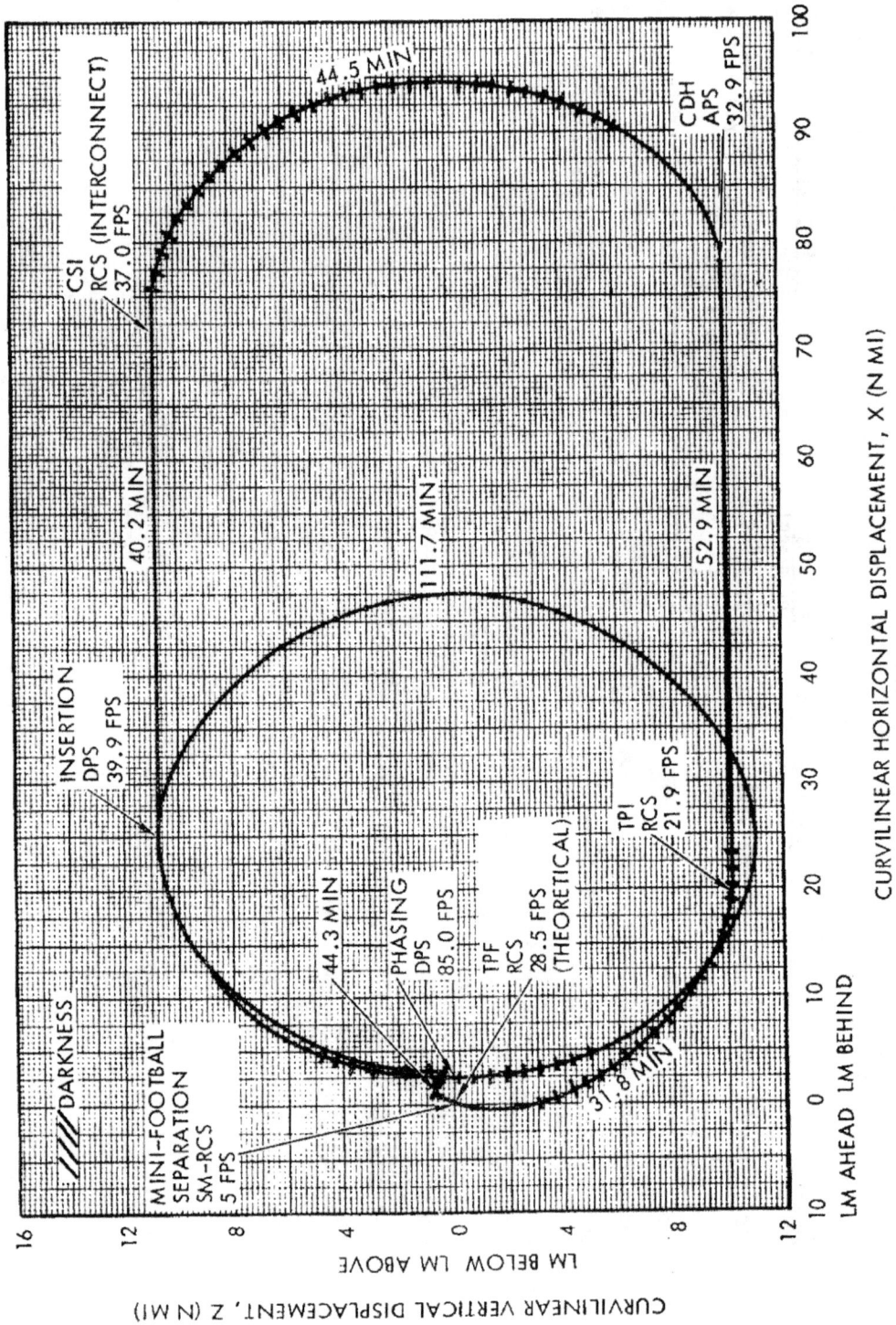

FLIGHT PLAN SUMMARY

Following is a brief summary of tasks to be accomplished in Apollo 9 on a day-to-day schedule. Apollo 9 work days are not on a 24-hour basis but rather on a variable mission phase and crew activity basis. Rest periods are scheduled at irregular intervals between mission phases.

Launch Day (0-19 hours elapsed time):

- CSM systems checkout following insertion into 103 nm orbit

- Preparations for transposition, docking and LM ejection

- CSM separation, transposition and docking with lunar module

- LM pressurization to equal of CSM

- LM ejection from spacecraft/LM adapter

- Spacecraft evasive maneuver to observe S-IVB restart

- Docked posigrade service propulsion system burn raises apogee to 128 nm

- Daylight star check and sextant calibration

Second day (19-40 hours GET):

- Second SPS docked burn tests digital autopilot stability at 40 per cent of full amplitude gimbal stroke posigrade, raises apogee to 192 nm

- Third SPS docked burn tests digital autopilot stability at full stroke; although mostly out-of-plane, burn raises apogee to 270 nm

- Fourth docked SPS burn, out-of-plane; orbit remains 115x270 nm

Third day (40-67 GET):

- Commander and LM pilot transfer to LM

- LM systems checkout

- LM alignment optical telescope daylight star visibility check

- LM S-Band steerable antenna check

-more-

* LM platform alignment with combination of known CSM attitudes and voiced data from ground

* LM RCS engines hot firing

* Docked LM descent engine burn: out of plane, no orbit change

● LM crew transfer back into command module

* Fifth SPS docked burn circularizes orbit to 133 nm

Fourth day (67-87 GET):

* Commander and LM pilot transfer to LM and begin preparations for EVA

* CSM and LM depressurized and hatches opened

● LM pilot leaves through LM front hatch, mounts camera; command module pilot mounts camera on command module open hatch

● LM pilot carries out two-hour EVA in transfer to command module and back to LM for period in "golden slippers" on LM porch; tests LM TV camera

* LM pilot enters LM, closes hatch; LM and command module re-pressurized; LM crew returns to command module

Fifth day (87-114 GET):

* Commander and LM pilot transfer to LM and prepare for un-docking

* LM checkout simulating preparations for lunar landing descent

* LM rendezvous radar self-test

* LM/CSM separation, command module pilot inspects and photo-graphs LM landing gear

* LM rendezvous radar lock-on with CSM transponder

* Descent engine phasing burn places LM in 119x145 nm orbit for rendezvous radial separation

● LM descent engine posigrade burn inserts LM into 142x144 nm orbit for coelliptic rendezvous maneuver

* LM RCS retrograde burn and staging of LM descent stage; LM ascent stage now in 120x139 nm orbit

* LM ascent engine coelliptic burn places LM in 119x121 nm orbit; LM orbit now constant about 10 nm below CSM

* LM RCS terminal phase initiation burn along line of sight toward CSM, followed by midcourse correction burns

* Rendezvous and docking

* LM crew transfer to command module

* LM jettisoned and ascent engine ground-commanded for burn to depletion

    Sixth day (114-139 GET):

* Sixth SPS burn lowers CSM perigee to 95 nm

    Seventh day (139-162 GET):

* Landmark tracking over U.S. and South Atlantic

    Eighth day (162-185 GET):

* Seventh SPS burn raises apogee to 210 nm

    Ninth day (185-208 GET):

* No major mission activities planned

    Tenth day (208-231 GET):

* No major mission activities planned

    Eleventh day (231 GET to splashdown):

* Entry preparations

* SPS fps retrograde deorbit burn

* CM/SM separation, entry and splashdown.

APOLLO 9 ALTERNATE MISSIONS

Any of the several alternate mission plans possible for Apollo 8 will focus upon meeting the most lunar module test objectives. Depending upon when in the mission timeline a failure occurs, and what the nature of the failure is, an alternate mission plan will be chosen in real time.

Failures which would require a shift to an alternate mission plan are cross-matched against mission timeline periods in an alternate mission matrix from which the flight control team in Mission Control Center would choose an alternate promising the maximum in mission objectives met.

The functional failure side of the alternate mission matrix has 14 possible failures. They are: Early S-IVB cutoff, forcing an SPS contingency orbit insertion (COI) and a CSM only alternate mission; LM cannot be ejected from the spacecraft/ LM adapter (SLA); SPS will not fire; problem with CSM lifetime; failure of either CSM coolant loop; unsafe descent stage; LM descent propulsion engine inoperable; extravehicular transfer takes longer than 15 minutes; loss of LM primary guidance, navigation and control system (PGNCS); loss of LM primary coolant loop; electrical power problem in LM descent stage; electrical problem in LM ascent stage; LM rendezvous radar failure; and loss of LM abort guidance system.

Seven basic alternate missions have been outlined, each of which has several possible subalternates stemming from when a failure occurs and how many of the mission's objectives have been accomplished.

The basic alternate missions are summarized as follows:

Alternate A: (No lunar module because of SPS contingency orbit insertion or failure of LM to eject from SLA.) Full-duration CSM only mission with the scheduled eight SPS burns on the nominal timeline.

Alternate B: (No SPS, lifetime problems on CSM and LM.) Would include transposition, docking and extraction of LM; LM systems evaluation, docked DPS burn, EVA, station keeping with ascent stage, long ascent engine burn, and RCS deorbit.

Alternate C: (Unsafe descent stage, EVA transfer runs overtime.) Crew would perform EVA after separating descent stage, the long ascent engineburn would be performed and the balance of the mission continue along the nominal timeline.

D Mission

ALTERNATE MISSIONS

ALTERNATE

DESCRIPTION

A. Contingency: LM fails or cannot be ejected from SLA.
    Perform CSM-only mission.

B. Contingency: Curtailed CSM systems performance.
    Accomplish priority objectives on accelerated time scale.

C. Contingency: DPS cannot be ignited, or fails during docked burn.
    Perform EVA and long APS burn. Accomplish CSM objectives.

D. Contingency: Loss of CSM ECS coolant loop.
    Depending on time of occurrence, conduct LM evaluation, execute docked DPS burn,
    station keeping, and long APS burn.

E. Contingency: Various LM subsystem failures.
    Perform rendezvous as modified in real-time.

F. Contingency: Loss of PGNCS.
    Delete docked DPS burn and long APS burn, perform EVA, and substitute CSM active
    rendezvous.

G. Contingency: LM primary coolant loop failure.
    Delete docked DPS burn. Perform EVA and long APS burn.

Alternate D: (CSM and/or LM lifetime problems, failure of either CSM coolant loop.) Mission plan would be reshaped to include transposition, docking and LM extraction, LM systems evaluation, docked descent engine burn, staging and long ascent engine burn, and continuation along nominal timeline.

Alternate E: (Unsafe descent stage, descent engine failure, LM primary coolant loop failure, electrical problems in either LM stage, PGNCS failure, rendezvous radar failure, or loss of LM abort guidance system.) Four possible modified rendezvous plans are in this alternate, each depending upon the nature of the systems failure and when it takes place. The modified rendezvous are: station keeping, mini-football rendezvous, football rendezvous and CSM active rendezvous.

Alternate F: (Failure of LM primary guidance, navigation and control system.) This alternate would delete docked descent engine burn, long ascent engine burn, but would include SPS burn No. 5, EVA station keeping and docking with LM ascent stage, and a CSM-active rendezvous.

Alternate G: (Failure of LM primary coolant loop or descent engine inoperable.) The docked descent engine burn and the LM-active rendezvous is deleted in this alternate, but the long ascent engine burn is done after undocking and the mission then follows the nominal timeline.

-more-

ABORT MODES

The Apollo 9 mission can be aborted at any time during the launch phase or during later phases after a successful insertion into Earth orbit.

Abort modes can be summarized as follows:

Launch phase --

Mode I - Launch escape tower propels command module away from launch vehicle. This mode is in effect from about T-20 minutes when LES is armed until LES jettison at 3:16 GET and command module landing point can range from the Launch Complex 39A area to 520 nm (600 sm, 964 km) downrange.

Mode II - Begins when LES is jettisoned and runs until the SPS can be used to insert the CSM into a safe orbit (9:22 GET) or until landing points threaten the African coast. Mode II requires manual separation, entry orientation and full-lift entry with landing between 400 and 3,200 nm (461-3,560 sm, 741-5,931 km) downrange.

Mode III - Begins when full-lift landing point reaches 3,200 nm (3,560 sm, 5,931 km) and extends through orbital insertion. The CSM would separate from the launch vehicle, and if necessary, an SPS retrograde burn would be made, and the command module would be flown half-lift to entry and landing at approximately 3,350 nm (3,852 sm, 6,197 km) downrange.

Mode IV and Apogee Kick - Begins after the point the SPS could be used to insert the CSM into an Earth parking orbit -- from about 9:22 GET. The SPS burn into orbit would be made two minutes after separation from the S-IVB and the mission would continue as an Earth orbit alternate. Mode IV is preferred over Mode III. A variation of Mode IV is the Apogee Kick in which the SPS would be ignited at first apogee to raise perigee for a safe orbit.

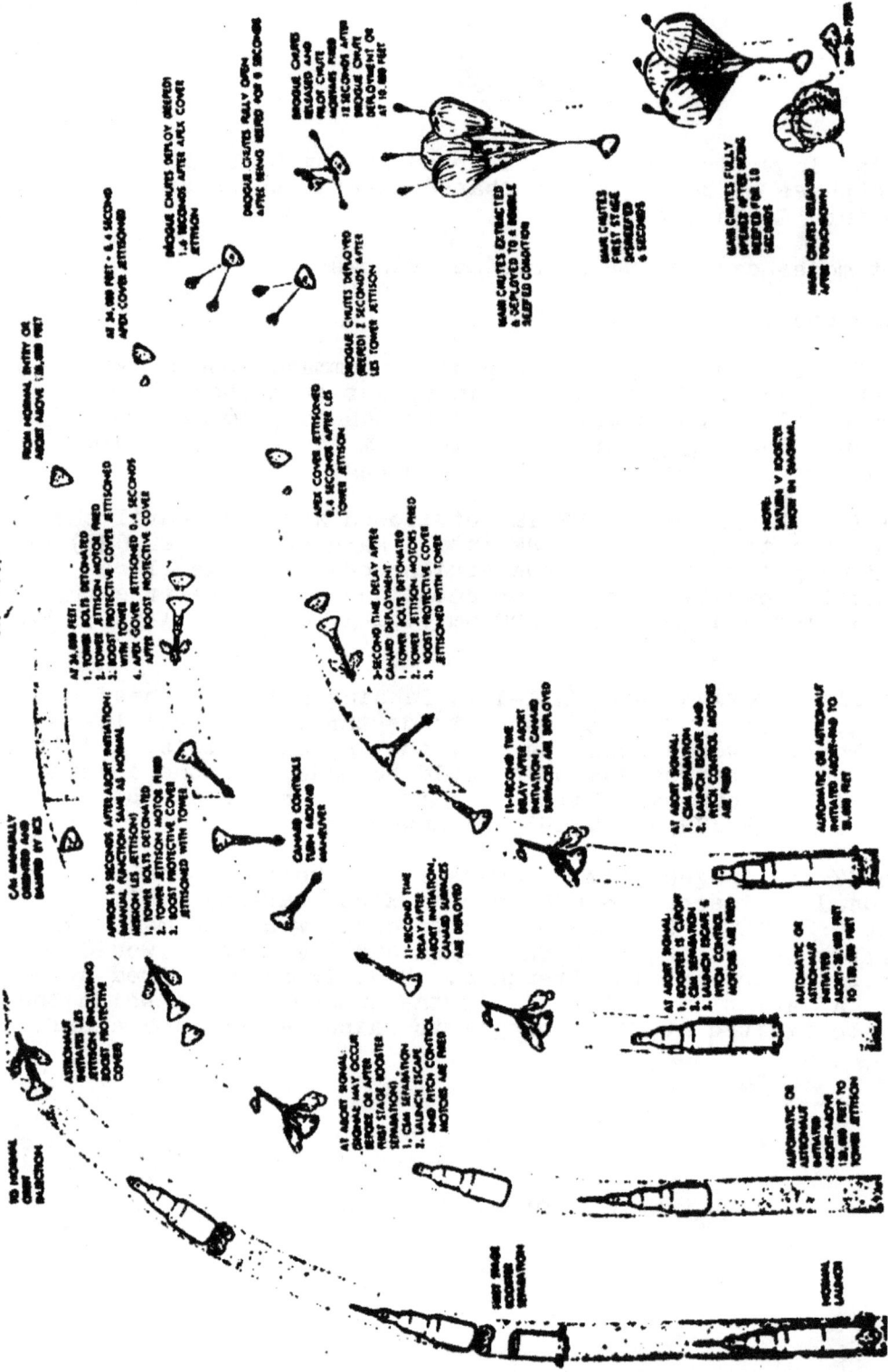

LES ABORT DIAGRAM

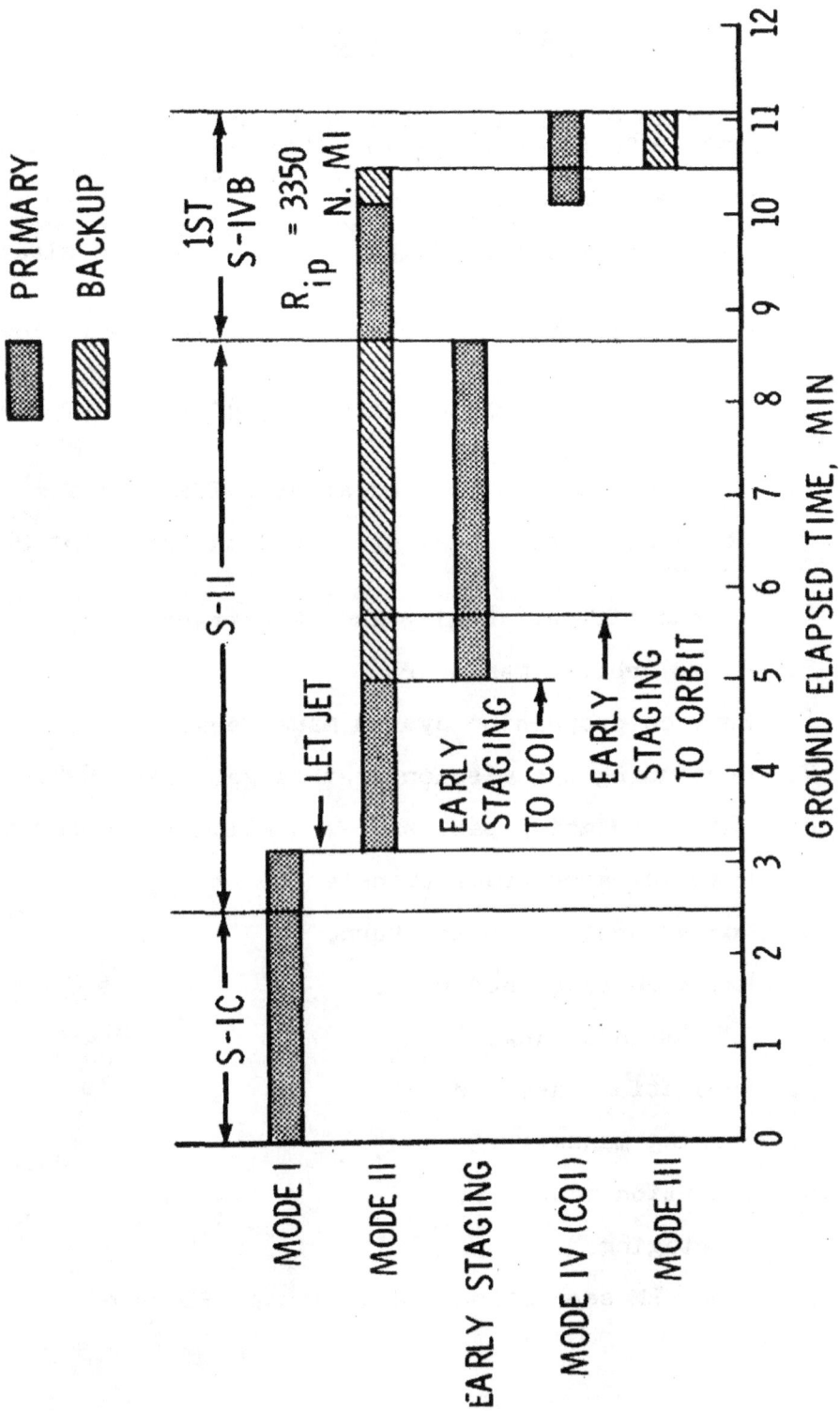

MISSION D LAUNCH ABORT TIMELINE

APOLLO 9 GO/NO-GO DECISION POINTS

Like Apollo 8, Apollo 9 will be flown on a step-by-step commit point or Go/No-Go basis in which the decision will be made prior to each major maneuver whether to continue the mission or to switch to one of the possible alternate missions. The Go/No-Go decisions will be based upon the joint opinions of the flight crew and the flight control teams in Mission Control Center.

Go/No-Go decisions will be made prior to the following events:

1. Launch phase Go/No-Go at 9 min. 40 sec. GET for orbit insertion.

2. S-IVB orbit coast period after S-IVB cutoff.

3. Continue mission past preferred target point 2-1 to target point 6-4.

4. Transposition, docking and LM extraction.

5. S-IVB orbital maneuvers.

6. Service propulsion system maneuvers.

7. Continuing the mission past target point 6-4.

8. Daily for going past the West Atlantic target point.

9. Crew intravehicular transfer to LM.

10. Docked descent engine burn.

11. Extravehicular activity.

12. CSM/LM undocking.

13. Separation maneuver.

14. Phasing maneuver.

15. Insertion maneuver.

16. LM staging.

17. Final LM separation and unmanned APS burn.

RECOVERY OPERATIONS

The primary landing point for Apollo 9 is in the West Atlantic at 59.9 degrees West Longitude by 30.1 degrees North Latitude for a nominal full-duration mission. Prime recovery vessel is the helicopter landing platform USS Guadalcanal.

Splashdown for a nominal mission launched on time at 11 a.m. EST, Feb. 28 will be at 9:46 a.m. EST, Mar. 10.

Other ships along the launch-phase ground track, in addition to the Guadalcanal, will be the Apollo instrumentation ship Vanguard and the destroyer USS Chilton off the west coast of Africa. Ships on station in Pacific contingency landing areas include one vessel in the West Pacific and two in the mid-Pacific.

In addition to surface vessels deployed in the launch abort area and the primary recovery vessel in the Atlantic, 16 HC-130 aircraft will be on standby at eight staging bases around the Earth: Tachikawa, Japan; Pago Pago, Samoa; Hawaii; Bermuda; Lajes, Azores; Ascension Island; Mauritius and Panama Canal Zone.

Apollo 9 recovery operations will be directed from the Recovery Operations Control Room in the Mission Control Center and will be supported by the Atlantic Recovery Control Center, Norfolk, Va.; Pacific Recovery Control Center, Kunia, Hawaii; and control centers at Ramstein, Germany and Albrook AFB, Canal Zone.

Following splashdown and crew and spacecraft recovery, the Guadalcanal will steam toward Norfolk, Va. The flight crew will be flown by helicopter to Norfolk the morning after recovery from whence they will fly to the Manned Spacecraft Center, Houston

The spacecraft will be taken off at Norfolk upon the Guadalcanal's arrival and undergo deactivation for approximately five days. It then will be flown aboard a C-133B aircraft to Long Beach, Calif., and thence trucked to the North American Rockwell Space Division plant in Downey, Calif., for postflight analysis.

PHOTOGRAPHIC EQUIPMENT

Apollo 9 will carry two 70mm standard and one superwide-angle Hasselblad still cameras and two 16 mm Maurer sequence cameras. Film magazines for specific mission photographic objectives are carried for each camera.

The Standard Hasselblad cameras are fitted with 80 mm f/2.8 to f/22 Zeiss Planar lenses, and the Superwide Angle Hasselblad is fitted with a 38mm f/4.5 to f/22 Zeiss Biogon lens. The Maurer sequence cameras have bayonet-mount 75mm f/2.5, 18mm f/2 and 5mm f/2 interchangeable lenses available to the crew.

Hasselblad shutter speeds are variable from 1 sec. to 1/500 sec., and sequence camera frame rates of 1,6,12 and 24 frames-per-second can be selected.

Film emulsions have been chosen for each specific photographic task. For example, a medium speed color reversal film will be used for recording docking, EVA and rendezvous and a high-speed color film will be used for command module and lunar module cabin interior photography.

Camera accessories carried aboard Apollo 9 include mounting brackets, right-angle mirror attachments, haze filter, an exposure-measuring spotmeter, a ringsight common to both types of camera, an EVA camera tether and a sequence camera remote controller for EVA photography. Power cables also are included.

APOLLO 9 EXPERIMENT

S065 Experiment -- Multispectral Photography

This experiment, being flown for the first time, is designed to obtain multispectral photography from space over selected land and ocean areas.

Equipment for the experiment consists of four Model 500-EL Hasselblad cameras operated by electric motors, installed in a common mount and synchronized for simultaneous exposure. The mount is installed in the command module hatch window during photographic operations and the spacecraft will be oriented to provide vertical photography. A manual introvolometer is used to obtain systematic overlapping (stereo) photography.

Each camera has a standard 80 milimeter focal length lens and a single film magazine containing from 160 to 200 frames.

Film-filter combinations for the cameras are similar to the ones presently planned for the Earth Resources Technology Satellite (ERTS-A) payload.

Present plans for Earth photography emphasize coverage of the Southwest U. S., where ground information is more readily available. Other areas of high interest include Mexico and Brazil. The domestic area of interest includes Tuscon, El Paso, Dallas/Ft. Worth and the Welaco Agricultural Experiment Station in Southwest Texas.

A photographic operations room will be maintained around the clock by the Manned Spacecraft Center's Earth Resources Division. Direct access to the mission controller will provide in-flight reprogramming of photographic coverage to help optimize photographic coverage by considering weather and operational conditions as the mission progresses.

A meeting will be held at the Manned Spacecraft Center as soon as possible after the mission to allow approximately 20 participating scientific investigators and participating user agency representatives to review the photography and get briefed on the details of the experiment. It is estimated that it will take approximately two weeks for the photographic technology laboratory to make the high quality film duplicates that will be provided to each participant at this meeting. The participants will be asked to provide a report within 90 days on their preliminary experimentation and these reports will be compiled and published by NASA.

Films and filters and spectral ranges for the experiment are:

1) Infrared Aerographic film with an 89B filter, 700 mu to 900 mu, to provide narrow-band infrared data for comparison with the responses obtained in the visible region of the electromagnetic spectrum by the other cameras in the experiment.

2) Color IR with Wratten 15 filter, 510 mu to 900 mu region, to take advantage of plant reflectance in the near IR and to provide for maximum differentiation between natural and cultural features.

3) Panatomic-X with 25A filter, 580 mu into the IR region, to provide imagery of value in differentiating various types of land use and in enhancing high contrast objects such as clouds.

4) Panatomic-X with 58 filter, 480 mu to 620 mu region, to provide for maximum penetration of lakes and coastal water bottom topography.

The principal investigator is Dr. Paul D. Lowman, Jr., of the NASA Goddard Space Flight Center, Greenbelt.

APOLLO 9 ONBOARD TELEVISION

A lunar television camera of the type that will transmit a video signal back to Earth during Apollo lunar landing missions will be stowed aboard LM-3.

Two television transmissions are planned for Apollo 9 -- one during the first manning and systems checkout of the LM, and the other during Schweickart's EVA. The first TV pass will be a test with the camera simply warmed up and passively transmitting during the systems checkout, and will last some seven minutes (46:27 - 46:34 GET) during a pass over the MILA tracking station.

After Schweickart has transferred EVA from the LM to the command module and back and is restrained by the "golden slippers" on the LM porch, McDivitt will pass the TV camera out to him for a 10-minute pass over the Goldstone and MILA stations (75:05 - 75:15 GET).

The video signal is transmitted to ground stations by the LM S-Band transmitter. Goldstone, Calif. and Merritt Island, Fla., are the two MSFN stations equipped for scan conversion and output to the Mission Control Center, although other MSFN stations are capable of recording the TV signal at the slow-scan rate.

The lunar television camera weighs 7.25 pounds and draws 6.5 watts of 24-32 volts DC power. Scan rate is 10 frames/sec. at 320 lines/frame. The camera body is 10.6 inches long, 6.5 inches wide and 3.4 inches deep. The bayonet lens mount permits lens changes by a crewman in a pressurized suit. Lenses for the camera include a lunar day lens, lunar night lens, a wide-angle lens and a 100mm telephoto lens. The wide-angle, and lunar day lens will be carried with the Apollo 9 camera.

A tubular fitting on the end of the electrical power cable which plugs into the bottom of the camera serves as a handgrip.

The Apollo lunar television camera is built by Westinghouse Electric Corporation Aerospace Division, Baltimore, Md. TV cameras carried on Apollo 7 and 8 were made by RCA.

COMMAND AND SERVICE MODULE STRUCTURE, SYSTEMS

The Apollo spacecraft for the Apollo 9 mission is comprised of a Command module 104, service module 104, lunar module 3, a spacecraft-lunar module adapter (SLA) 12 and a launch escape system. The SLA serves as a mating structure between the instrument unit atop the S-IVB stage of the Saturn V launch vehicle and as a housing for the lunar module.

Launch Escape System (LES)--Propels command module to safety in an aborted launch. It is made up of an open-frame tower structure mounted to the command module by four frangible bolts, and three solid-propellant rocket motors : a 147,000 pound-thrust launch escape system motor, a 2,400-pound-thrust pitch control motor and a 31,500-pound-thrust tower jettison motor. Two canard vanes near the top deploy to turn the command module aerodynamically to an attitude with the heat-shield forward. Attached to the base of the launch escape tower is a boost protective cover composed of glass, cloth and honeycomb, that protects the command module from rocket exhaust gases from the main and the jettison motor. The system is 33 feet tall, four feet in diameter at the base and weighs 8,848 pounds.

Command Module (CM) Structure--The basic structure of the command module is a pressure vessel encased in heat-shields, cone-shaped 12 feet high, base diameter of 12 feet 10 inches, and launch weight 12,405 pounds.

The command module consists of the forward compartment which contains two negative pitch reaction control engines and components of the Earth landing system; the crew compartment, or inner pressure vessel, containing crew accommodations, controls and displays, and spacecraft systems; and the aft compartment housing ten reaction control engines and propellant tankage.

Heat-shields around the three compartments are made of brazed stainless steel honeycomb with an outer layer of phenolic epoxy resin as an ablative material. Heat-shield thickness, varying according to heat loads, ranges from 0.7 inches (at the apex) to 2.7 inches on the aft side.

The spacecraft inner structure is of aluminum alloy sheet-aluminum honeycomb bonded sandwich ranging in thickness from 0.25 inches thick at forward access tunnel to 1.5 inches thick at base.

CSM 104 and LM-3 will carry for the first time the probe-and-drogue docking hardware. The probe assembly is a folding coupling and impact attentuating device mounted on the CM tunnel that mates with a conical drogue mounted on the LM docking tunnel. After the docking latches are dogged down following a docking maneuver, both the probe and drogue assemblies are removed from the vehicle tunnels and stowed to allow free crew transfer between the CSM and LM.

Q-BALL (NOSE CONE)

PITCH CONTROL MOTOR

CANARDS

JETTISON MOTOR

LAUNCH ESCAPE MOTOR

STRUCTURAL SKIRT

LAUNCH ESCAPE TOWER

TOWER ATTACHMENT (4)

COMMAND MODULE

BOOST PROTECTIVE COVER

EPS RADIATOR

REACTION CONTROL SYSTEM ENGINES

SERVICE MODULE

ECS RADIATOR

SPS ENGINE EXPANSION NOZZLE

SPACECRAFT LM ADAPTER (SLA)

SLA PANEL JUNCTION (BETWEEN FWD AND AFT PANELS)

INSTRUMENT UNIT (SHOWN AS REFERENCE)

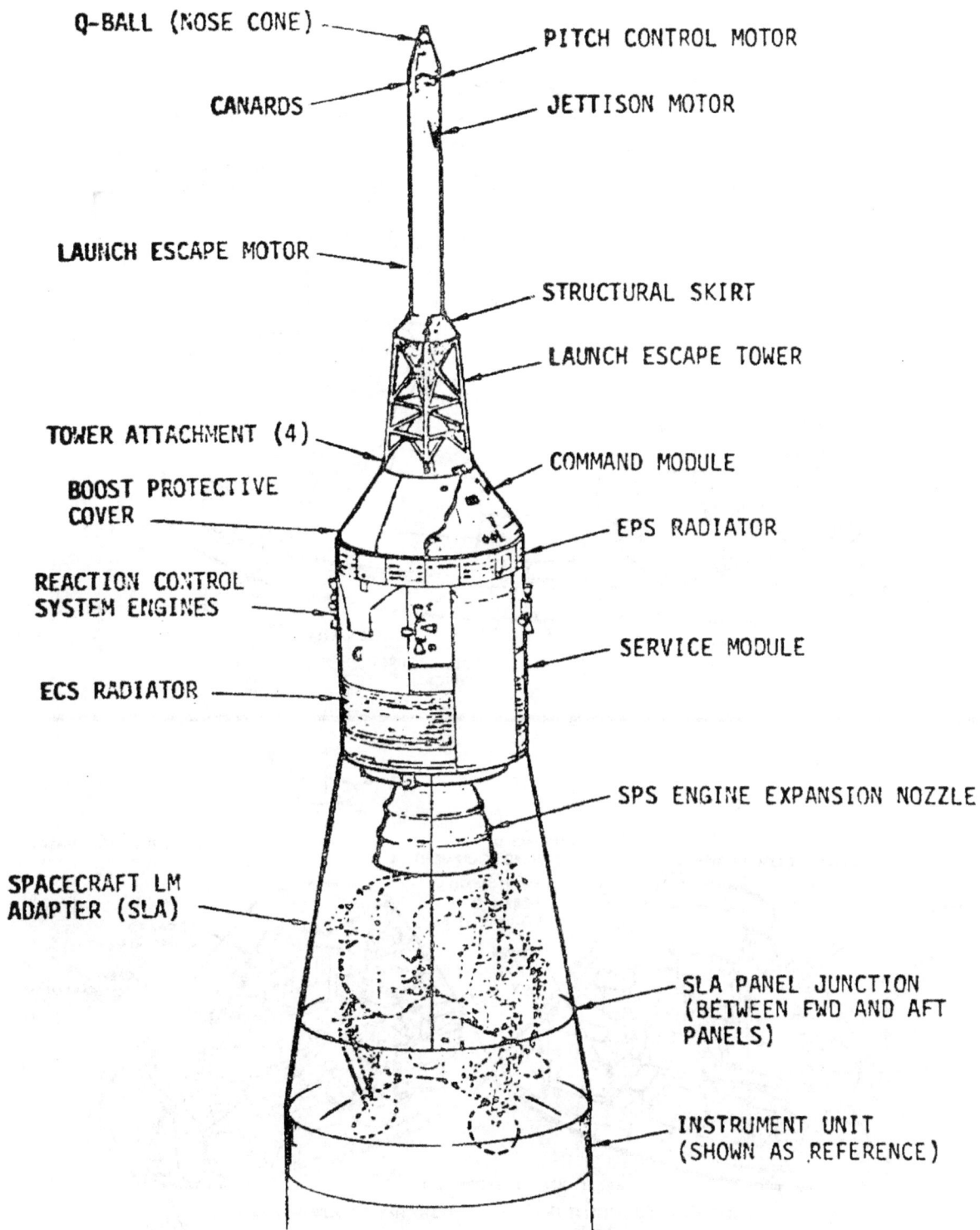

SPACECRAFT CONFIGURATION

-more-

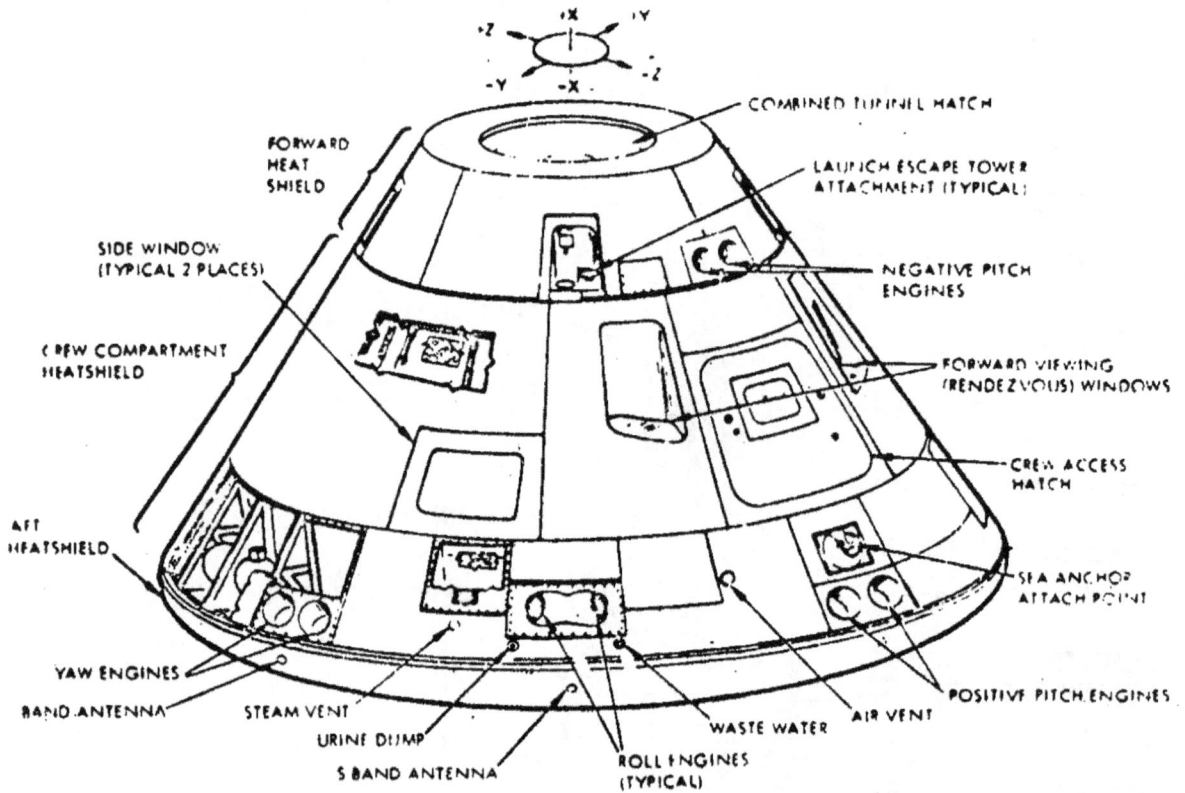

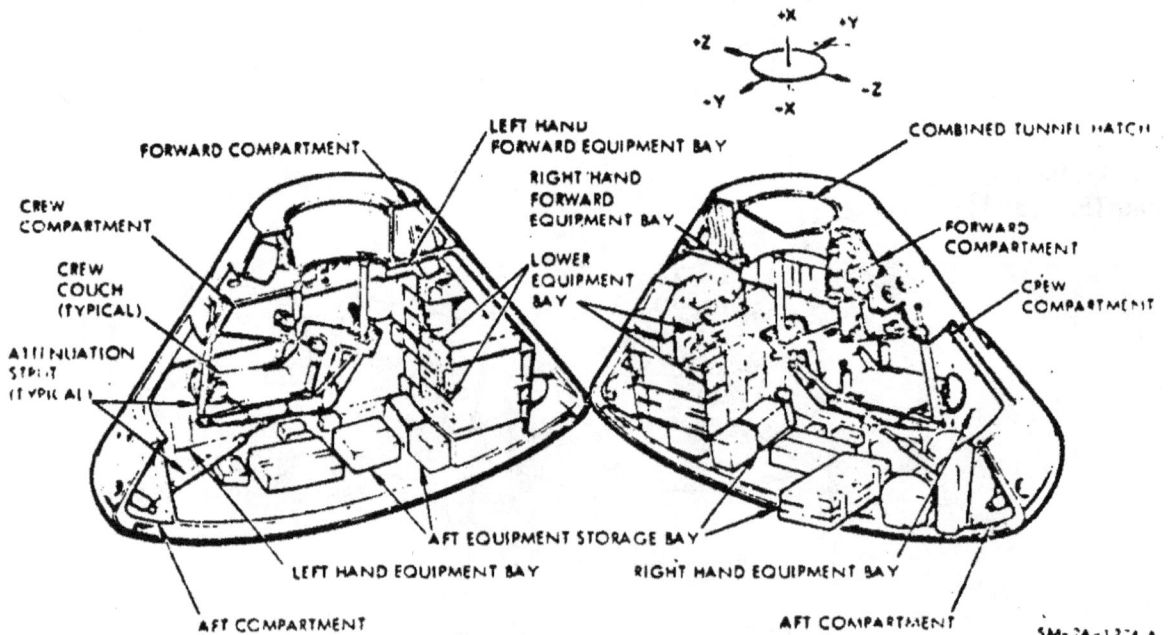

COMBINED TUNNEL HATCH

LAUNCH ESCAPE TOWER
ATTACHMENT (TYPICAL)

NEGATIVE PITCH
ENGINES

FORWARD VIEWING
(RENDEZVOUS) WINDOWS

CREW ACCESS
HATCH

SEA ANCHOR
ATTACH POINT

FORWARD
HEAT
SHIELD

SIDE WINDOW
(TYPICAL 2 PLACES)

CREW COMPARTMENT
HEATSHIELD

AFT
HEATSHIELD

YAW ENGINES

BAND ANTENNA

STEAM VENT

URINE DUMP

S BAND ANTENNA

ROLL ENGINES
(TYPICAL)

WASTE WATER

AIR VENT

POSITIVE PITCH ENGINES

FORWARD COMPARTMENT

LEFT HAND
FORWARD EQUIPMENT BAY

RIGHT HAND
FORWARD
EQUIPMENT BAY

LOWER
EQUIPMENT
BAY

COMBINED TUNNEL HATCH

FORWARD
COMPARTMENT

CREW
COMPARTMENT

CREW
COMPARTMENT

CREW
COUCH
(TYPICAL)

ATTENUATION
STRUT
(TYPICAL)

AFT EQUIPMENT STORAGE BAY

LEFT HAND EQUIPMENT BAY

RIGHT HAND EQUIPMENT BAY

AFT COMPARTMENT

AFT COMPARTMENT

SM-2A-1274 A

Service Module (SM) Structure--The service module is a cylinder 12 feet 10 inches in diameter by 22 feet long. For the Apollo 9 mission, it will weigh 36,159 pounds (16,416.2 kg) at launch. Aluminum honeycomb panes one inch thick form the outer skin, and milled aluminum radial beams separate the interior into six sections containing service propulsion system and reaction control fuel-oxidizer tankage, fuel cells, cryogenic oxygen and hydrogen, and onboard consumables.

Spacecraft-LM Adapter (SLA) Structure--The spacecraft LM adapter is a truncated cone 28 feet long tapering from 260 inches diameter at the base to 154 inches at the forward end at the service module mating line. Aluminum honeycomb 1.75 inches thick is the stressed-skin structure for the spacecraft adapter. The SLA weighs 4,107 pounds.

CSM Systems

Guidance, Navigation and Control System (GNCS)--Measures and controls spacecraft position, attitude and velocity, calculates trajectory, controls spacecraft propulsion system thrust vector and displays abort data. The Guidance System consists of three subsystems: inertial, made up of an inertial measuring unit and associated power and data components; computer which processes information to or from other components; and optics, including scanning telescope, sextant for celestial and/or landmark spacecraft navigation.

Stabilization and Control System (SCS) --Controls spacecraft rotation, translation and thrust vector and provides displays for crew-initiated maneuvers; backs up the guidance system. It has three subsystems; attitude reference, attitude control and thrust vector control.

Service Propulsion System (SPS)--Provides thrust for large spacecraft velocity changes through a gimbal-mounted 20,500-pound-thrust hypergolic engine using nitrogen tetroxide oxidizer and a 50-50 mixture of unsymmetrical dimethyl hydrazine and hydrazine fuel. Tankage of this system is in the service module. The system responds to automatic firing commands from the guidance and navigation system or to manual commands from the crew. The engine provides a constant thrust rate. The stabilization and control system gimbals the engine to fire through the spacecraft center of gravity.

Reaction Control System (RCS)--The Command Module and the Service Module each has its own independent system, the CM RCS and the SM RCS respectively. The SM RCS has four identical RCS "quads" mounted around the SM 90 degrees apart. Each quad has four 100 pound-thrust engines, two fuel and two oxidizer tanks and a helium pressurization sphere. The SM RCS provides redundant spacecraft attitude control through cross-coupling logic inputs from the Stabilization and Guidance Systems.

SERVICE MODULE
BLOCK II

EPS RADIATORS

SM—RCS

ECS RADIATOR

—Z

—Y

SPS

DOCKING LIGHTS

CM/SM FAIRING

S—BAND HIGH—GAIN ANTENNA

+Z

RADIAL BEAM TRUSS 16 PLACES

FAIRING

FUEL CELL POWER PLANTS

O2 TANKS

H2 TANKS

FUEL SUMP TANK

SPS ENGINE EXPANSION NOZZLE

OXIDIZER SUMP TANK

EPS RADIATOR

SPS HELIUM TANKS

RCS QUAD

ECS SPACE RADIATOR

FUEL STORAGE TANK

FUEL FILL POINT

—Z

—Y

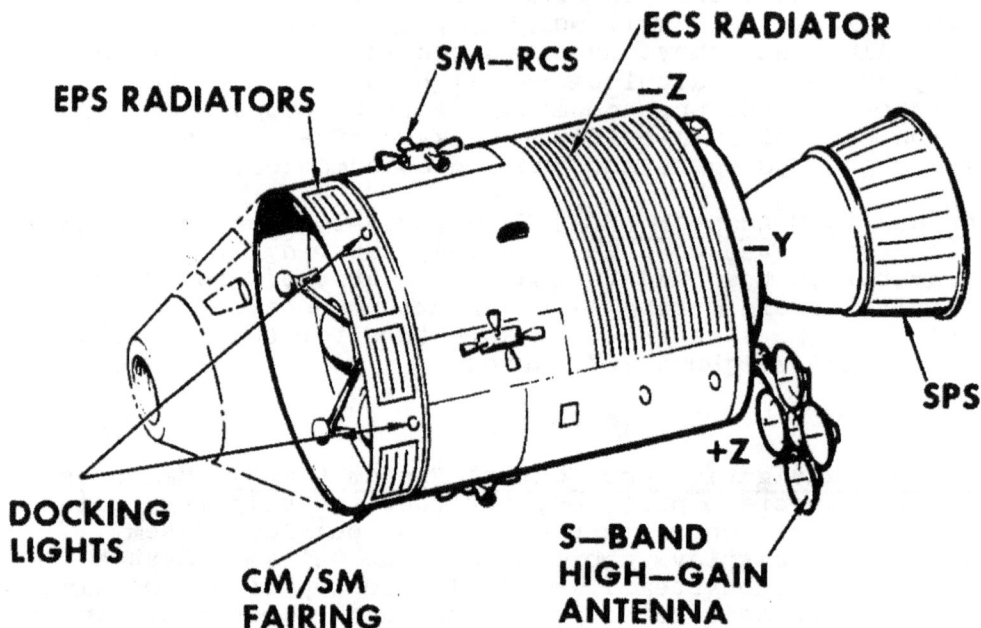

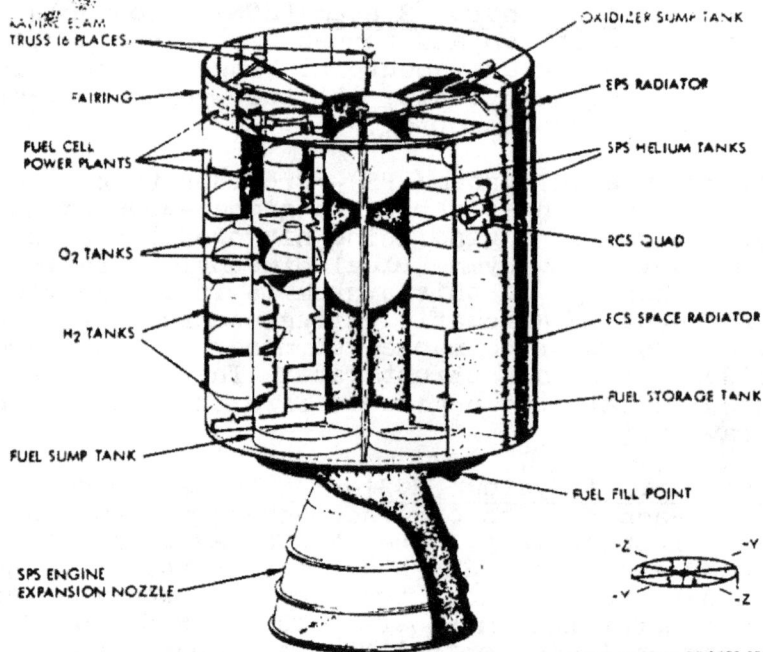

1 AND 4 ARE 50-DEGREE SECTORS
2 AND 5 ARE 70-DEGREE SECTORS
3 AND 6 ARE 60-DEGREE SECTORS

Small velocity change maneuvers can also be made with the SM RCS. The CM RCS consists of two independent six-engine subsystems of six 94 pounds-thrust engines each. Both subsystems are activated after separation from the SM: one is used for spacecraft attitude control during entry. The other serves in standby as a backup. Propellants for both CM and SM RCS are monomethyl hydrazine fuel and nitrogen tetroxide oxidizer with helium pressurization. These propellants are hypergolic, i.e., they burn spontaneously when combined without need for an igniter.

Electrical Power System (EPS)--Consists of three, 31-cell Bacon-type hydrogen-oxygen fuel cell power plants in the service module which supply 28-volt DC power, three 28-volt DC zinc-silver oxide main storage batteries in the command module lower equipment bay, and three 115-200-volt 400 hertz three-phase AC inverters powered by the main 28-volt DC bus. The inverters are also located in the lower equipment bay. Cryogenic hydrogen and oxygen react in the fuel cell stacks to provide electrical power, potable water and heat. The command module main batteries can be switched to fire pyrotechnics in an emergency. A battery charger restores selected batteries to full strength as required with power from the fuel cells.

Environmental Control System (ECS)--Controls spacecraft atmosphere, pressure and temperature and manages water. In addition to regulating cabin and suit gas pressure, temperature and humidity, the system removes carbon dioxide, odors and particles, and ventilates the cabin after landing. It collects and stores fuel cell potable water for crew use, supplies water to the glycol evaporators for cooling, and dumps surplus water overboard through the urine dump valve. Proper operating temperature of electronics and electrical equipment is maintained by this system through the use of the cabin heat exchangers, the space radiators and the glycol evaporators.

Telecommunications System--Provides voice, television telemetry and command data and tracking and ranging between the spacecraft and earth, between the command module and the lunar module and between the spacecraft and the extravehicular astronaut. It also provides intercommunications between astronauts. The telecommunications system consists of pulse code modulated telemetry for relaying to Manned Space Flight Network stations data on spacecraft systems and crew condition, VHF/AM voice, and unified S-Band tracking transponder, air-to-ground voice communications, onboard television (not installed on CM 104) and a VHF recovery beacon. Network stations can transmit to the spacecraft such items as updates to the Apollo guidance computer and central timing equipment, and real-time commands for certain onboard functions. More than 300 CSM measurements will be telemetered to the MSFN.

- more -

The high-gain steerable S-Band antenna consists of four,
31-inch-diameter parabolic dishes mounted on a folding boom at
the aft end of the service module. Nested alongside the service
propulsion system engine nozzle until deployment, the antenna
swings out at right angles to the spacecraft longitudinal axis,
with the boom pointing 52 degrees below the heads-up horizontal.
Signals from the ground stations can be tracked either automatically or manually with the antenna's gimballing system. Normal
S-Band voice and uplink/downlink communications will be handled
by the omni and high-gain antennas.

Sequential System--Interfaces with other spacecraft systems
and subsystems to initiate time critical functions during launch,
docking maneuvers, pre-orbital aborts and entry portions of a
mission. The system also controls routine spacecraft sequencing
such as service module separation and deployment of the Earth
landing system.

Emergency Detection System (EDS)--Detects and displays to
the crew launch vehicle emergency conditions, such as excessive
pitch or roll rates or two engines out, and automatically or
manually shuts down the booster and activates the launch escape
system; functions until the spacecraft is in orbit.

Earth Landing System (ELS)--Includes the drogue and main
parachute system as well as post-landing recovery aids. In a
normal entry descent, the command module forward heat shield
is jettisoned at 24,000 feet, permitting mortar deployment of
two reefed 16.5-foot diameter drogue parachutes for orienting
and decelerating the spacecraft. After disreef and drogue release, three pilot mortar deployed chutes pull out the three
main 83.3-foot diameter parachutes with two-stage reefing to
provide gradual inflation in three steps. Two main parachutes
out of three can provide a safe landing.

Recovery aids include the uprighting system, swimmer inter-
phone connections, sea dye marker, flashing beacon, VHF recovery
beacon and VHF transceiver. The uprighting systems consists of
three compressor-inflated bags to upright the spacecraft if it
should land in the water apex down (Stable II position).

Caution and Warning System--Monitors spacecraft systems for
out-of-tolerance conditions and alerts crew by visual and audible
alarms so that crewmen may trouble-shoot the problem.

Controls and Displays--Provide readouts and control functions
of all other spacecraft systems in the command and service modules.
All controls are designed to be operated by crewmen in pressurized
suits. Displays are grouped by system according to the frequency
the crew refers to them.

- more -

LUNAR MODULE STRUCTURES, SYSTEMS

The lunar module is a two-stage vehicle designed for space operations near and on the Moon or in Earth orbit developmental missions such as Apollo 9. The LM is incapable of reentering the atmosphere and is, in effect, a true spacecraft.

Joined by four explosive bolts and umbilicals, the ascent and descent stages of the LM operate as a unit until staging, when the ascent stage functions as a single spacecraft for rendezvous and docking with the CSM.

Three main sections make up the ascent stage: the crew compartment, midsection and aft equipment bay. Only the crew compartment and midsection can be pressurized (4.8 psig; 337.4 gm/sq cm) as part of the LM cabin; all other sections of the LM are unpressurized. The cabin volume is 235 cubic feet (6.7 cubic meters).

Structurally, the ascent stage has six substructural areas: crew compartment, midsection, aft equipment bay, thrust chamber assembly cluster supports, antenna supports and thermal and micrometeoroid shield.

The cylindrical crew compartment is a semimonocoque structure of machined longerons and fusion-welded aluminum sheet and is 92 inches (2.35 m) in diameter and 42 inches (1.07 m) deep. Two flight stations are equipped with control and display panels, armrests, body restraints, landing aids, two front windows, an overhead docking window and an alignment optical telescope in the center between the two flight stations.

Two triangular front windows and the 32-inch (.81 m) square inward-opening forward hatch are in the crew compartment front face.

External structural beams support the crew compartment and serve to support the lower interstage mounts at their lower ends. Ring-stiffened semimonocoque construction is employed in the midsection, with chem-milled aluminum skin over fusion-welded longerons and stiffeners. Fore-and-aft beams across the top of the midsection join with those running across the top of the cabin to take all ascent stage stress loads and, in effect, isolate the cabin from stresses.

The ascent stage engine compartment is formed by two beams running across the lower midsection deck and mated to the fore and aft bulkheads. Systems located in the midsection include the LM guidance computer, the power and servo assembly, ascent engine propellant tanks, RCS propellant tanks, the environmental control system, and the waste management section.

-more-

A tunnel ring atop the ascent stage meshes with the command module latch assemblies. During docking, the ring and clamps are aligned by the LM drogue and the CSM probe.

The docking tunnel extends downward into the midsection 16 inches (40 cm). The tunnel is 32 inches (.81 cm) in diameter and is used for crew transfer between the CSM and LM by crewmen in either pressurized or unpressurized extravehicular mobility units (EMU). The upper hatch on the inboard end of the docking tunnel hinges downward and cannot be opened with the LM pressurized.

A thermal and micrometeoroid shield of multiple layers of mylar and a single thickness of thin aluminum skin encases the entire ascent stage structure.

The descent stage consits of a cruciform load-carrying structure of two pairs of parallel beams, upper and lower decks, and enclosure bulkheads -- all of conventional skin-and-stringer aluminum alloy construction. The center compartment houses the descent engine, and descent propellant tanks are housed in the four square bays around the engine.

Four-legged truss outriggers mounted on the ends of each pair of beams serve as SLA attach points and as "knees" for the landing gear main struts.

Triangular bays between the main beams are enclosed into quadrants housing such components as the ECS water tank, helium tanks, descent engine control assembly of the guidance, navigation and control subsystem, ECS gaseous oxygen tank and batteries for the electrical power system. Like the ascent stage, the descent stage is encased in a mylar and aluminum alloy thermal and micrometeoroid shield.

The LM external platform, or "porch," is mounted on the forward outrigger just below the forward hatch. A ladder extends down the forward landing gear strut from the porch for crew lunar surface operations. Foot restraints ("golden slippers") have been attached to the LM-3 porch to assist the lunar module pilot during EVA photography. The restraints face the LM hatch.

In a retracted position until after the crew mans the LM, the landing gear struts are explosively extended to provide lunar surface landing impact attenuation. The main struts are filled with crushable aluminum honeycomb for absorbing compression loads. Footpads 37 inches (.95 m) in diameter at the end of each landing gear provide vehicle "flotation" on the lunar surface.

-more-

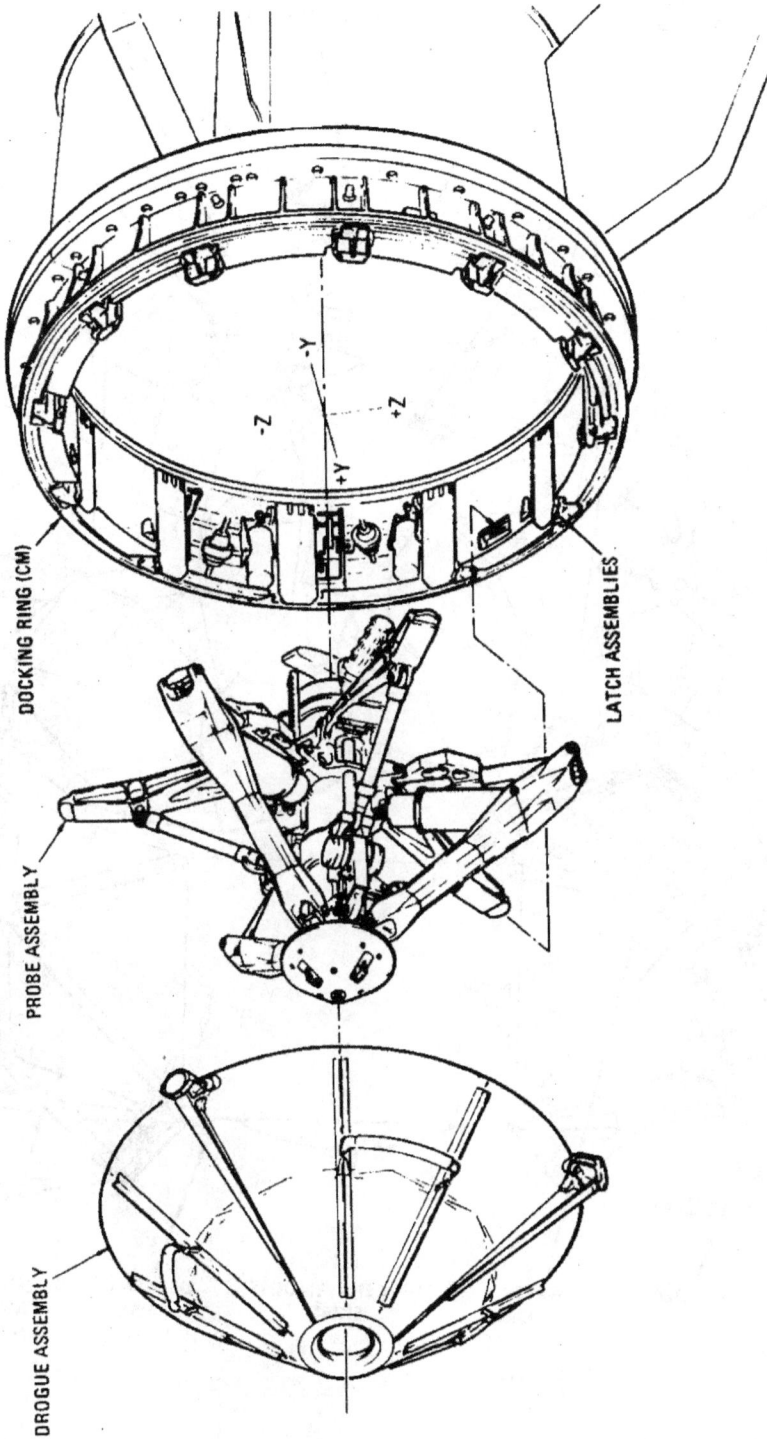

DOCKING RING (CM)

LATCH ASSEMBLIES

PROBE ASSEMBLY

DROGUE ASSEMBLY

-Y

-Z

+Z

+Y

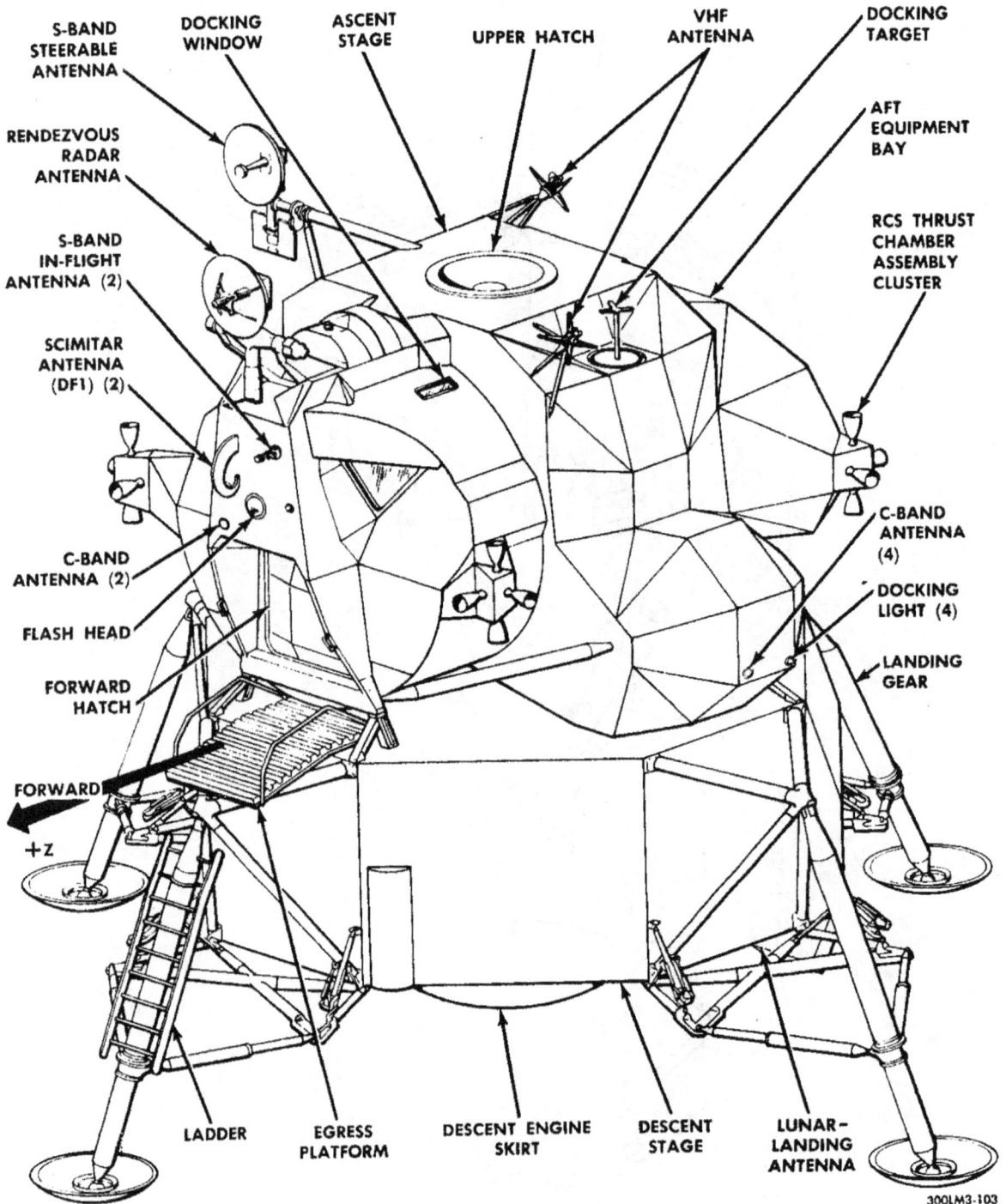

S-BAND
STEERABLE
ANTENNA

DOCKING
WINDOW

ASCENT
STAGE

UPPER HATCH

VHF
ANTENNA

DOCKING
TARGET

RENDEZVOUS
RADAR
ANTENNA

AFT
EQUIPMENT
BAY

S-BAND
IN-FLIGHT
ANTENNA (2)

RCS THRUST
CHAMBER
ASSEMBLY
CLUSTER

SCIMITAR
ANTENNA
(DF1) (2)

C-BAND
ANTENNA
(4)

C-BAND
ANTENNA (2)

DOCKING
LIGHT (4)

FLASH HEAD

LANDING
GEAR

FORWARD
HATCH

FORWARD

+Z

LADDER

EGRESS
PLATFORM

DESCENT ENGINE
SKIRT

DESCENT
STAGE

LUNAR-
LANDING
ANTENNA

300LM3-103

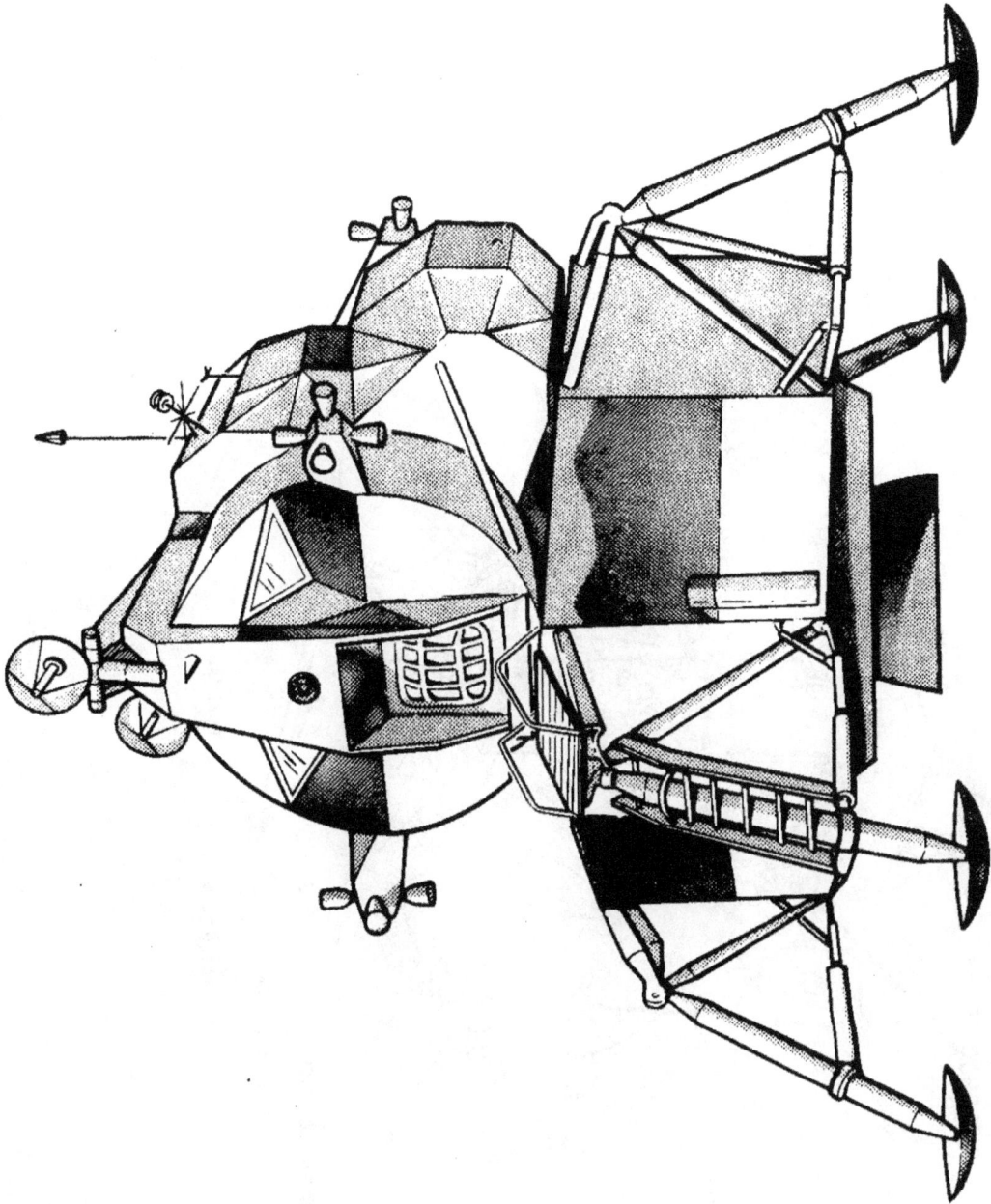

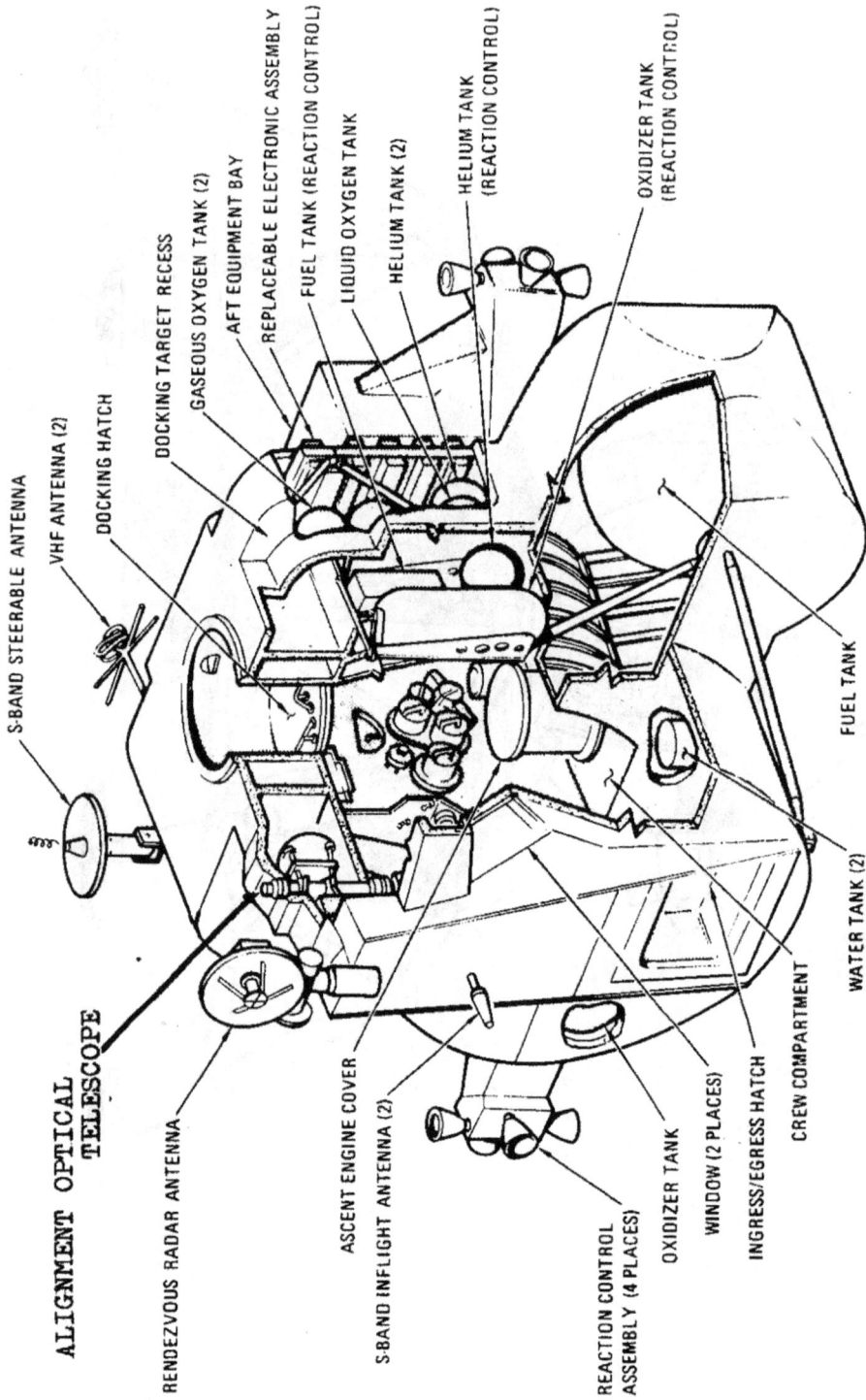

ALIGNMENT OPTICAL TELESCOPE

S-BAND STEERABLE ANTENNA

VHF ANTENNA (2)

DOCKING HATCH

DOCKING TARGET RECESS

GASEOUS OXYGEN TANK (2)

AFT EQUIPMENT BAY

REPLACEABLE ELECTRONIC ASSEMBLY

FUEL TANK (REACTION CONTROL)

LIQUID OXYGEN TANK

HELIUM TANK (2)

HELIUM TANK (REACTION CONTROL)

OXIDIZER TANK (REACTION CONTROL)

FUEL TANK

WATER TANK (2)

CREW COMPARTMENT

INGRESS/EGRESS HATCH

WINDOW (2 PLACES)

OXIDIZER TANK

REACTION CONTROL ASSEMBLY (4 PLACES)

S-BAND INFLIGHT ANTENNA (2)

ASCENT ENGINE COVER

RENDEZVOUS RADAR ANTENNA

P-74

Each pad is fitted with a lunar-surface sensing probe which signal the crew to shut down the descent engine upon contact with the lunar surface.

LM-3 flown on the Apollo 9 mission will have a launch weight of 32,000 pounds (14,507.8 kg). The weight breakdown is as follows:

Ascent stage, dry	5,071 lbs
Descent stage, dry	4,265 lbs
RCS propellants	605 lbs
DPS propellants	17,944 lbs
APS propellants	4,136 lbs
	32,021 lbs

LM-3 Spacecraft Systems

Electrical Power System -- The LM DC electrical system consists of six silver zinc primary batteries -- four in the descent stage and two in the ascent stage, each with its own electrical control assembly (ECA). Power feeders from all primary batteries pass through circuit breakers to energize the LM DC buses, from which 28-volt DC power is distributed through circuit breakers to all LM systems. AC power (117v 400Hz) is supplied by two inverters, either of which can supply spacecraft AC load needs to the AC buses.

Environmental Control System -- Consists of the atmosphere revitalization section, oxygen supply and cabin pressure control section, water management, heat transport section and outlets for oxygen and water servicing of the Portable Life Support System (PLSS).

Components of the atmosphere revitalization section are the suit circuit assembly which cools and ventilates the pressure garments, reduces carbon dioxide levels, removes odors and noxious gases and excessive moisture; the cabin recirculation assembly which ventilates and controls cabin atmosphere temperatures; and the steam flex duct which vents to space steam from the suit circuit water evaporator.

The oxygen supply and cabin pressure section supplies gaseous oxygen to the atmosphere revitalization section for maintaining suit and cabin pressure. The descent stage oxygen supply provides descent phase and lunar stay oxygen needs, and the ascent stage oxygen supply provides oxygen needs for the ascent and rendezvous phase.

-more-

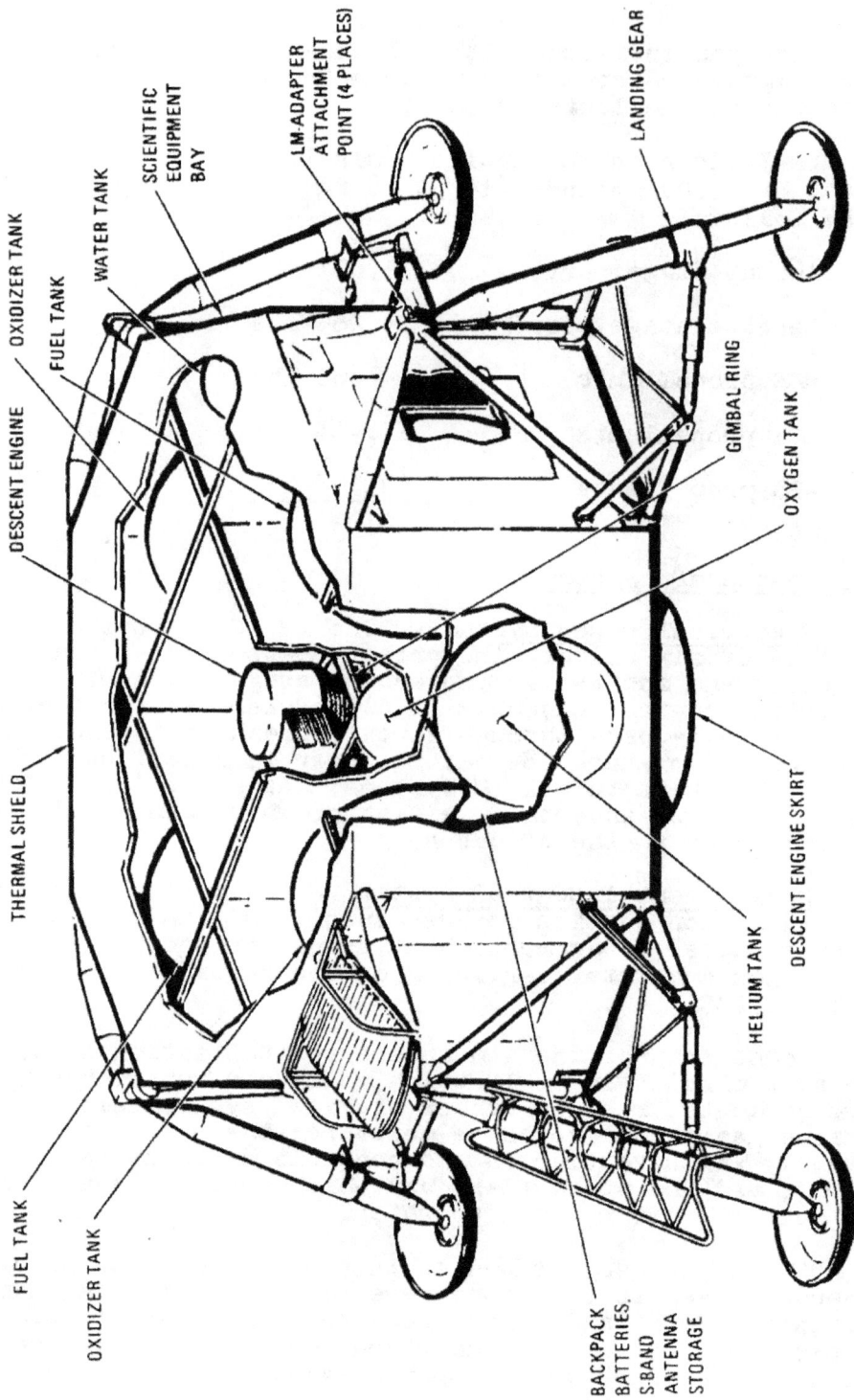

P-75

Water for drinking, cooling, firefighting and food preparation and refilling the PLSS cooling water servicing tank is supplied by the water management section. The water is contained in three nitrogen-pressurized bladder-type tanks, one of 367-pound capacity in the descent stage and two of 47.5-pound capacity in the ascent stage.

The heat transport section has primary and secondary water-glycol solution coolant loops. The primary coolant loop circulates water-glycol for temperature control of cabin and suit circuit oxygen and for thermal control of batteries and electronic components mounted on cold plates and rails. If the primary loop becomes inoperative, the secondary loop circulates coolant through the rails and cold plates only. Suit circuit cooling during secondary coolant loop operation is provided by the suit loop water boiler. Waste heat from both loops is vented overboard by water evaporation, or sublimators.

Communication System -- Two S-Band transmitter-receivers, two VHF transmitter-receivers, a UHF command receiver, a signal processing assembly and associated spacecraft antenna make up the LM communications system. The system transmits and receives voice, tracking and ranging data, and transmits telemetry data on 281 measurements and TV signals to the ground. Voice communications between the LM and ground stations is by S-Band, and between the LM and CSM voice is on VHF. In Earth orbital operations such as Apollo 9, VHF voice communications between the LM and the ground are possible. Developmental flight instrumentation (DFI) telemetry data are transmitted to MSFN stations by five VHF transmitters. Two C-Band beacons augment the S-Band system for orbital tracking.

The UHF receiver accepts command signals which are fed to the LM guidance computer for ground updates of maneuvering and navigation programs. The UHF receiver is also used to receive real-time commands which are on LM-3 to arm and fire the ascent propulsion system for the unmanned APS depletion burn. The UHF receiver will be replaced by an S-Band command system on LM-4 and subsequent spacecraft.

The Data Storage Electronics Assembly (DSEA) is a four-channel voice recorder with timing signals with a 10-hour recording capacity which will be brought back into the CSM for return to Earth. DSEA recordings cannot be "dumped" to ground stations.

LM antennas are one 26-inch diameter parabolic S-Band steerable antenna, two S-Band inflight antennas, two VHF inflight antennas, four C-Band antennas, and two UHF/VHF command/ DFI scimitar antennas.

Guidance, Navigation and Control System -- Comprised of six sections: primary guidance and navigation section (PGNS), abort guidance section (AGS), radar section, control electronics section (CES), and orbital rate drive electronics for Apollo and LM (ORDEAL).

* The PGNS is an inertial system aided by the alignment optical telescope, an inertial measurement unit, and the rendezvous and landing radars. The system provides inertial reference data for computations, produces inertial alignment reference by feeding optical sighting data into the LM guidance computer, displays position and velocity data, computes LM-CSM rendezvous data from radar inputs, controls attitude and thrust to maintain desired LM trajectory and controls descent engine throttling and gimbaling.

* The AGS is an independent backup system for the PGNS, having its own inertial sensor and computer.

* The radar section is made up of the rendezvous radar which provides and feeds CSM range and range rate, and line-of-sight angles for maneuver computation to the LM guidance computer; the landing radar which provides and feeds altitude and velocity data to the LM guidance computer during lunar landing. On LM-3, the landing radar will be in a self-test mode only. The rendezvous radar has an operating range from 80 feet to 400 nautical miles.

* The CES controls LM attitude and translation about all axes. Also controls by PGNS command the automatic operation of the ascent and descent engines, and the reaction control thrusters. Manual attitude controller and thrust-translation controller commands are also handled by the CES.

* ORDEAL displays on the flight director attitude indicator the computed local vertical in the pitch axis during circular Earth or lunar orbits.

Reaction Control System -- The LM has four RCS engine clusters of four 100-pound (45.4 kg) thrust engines each which use helium-pressurized hypergolic propellants. The oxidizer is nitrogen tetroxide, fuel is Aerozine 50 (50/50 hydrazine and unsymmetrical dimethyl hydrazine). Propellant plumbing, valves and pressurizing components are in two parallel, independent systems, each feeding half the engines in each cluster. Either system is capable of maintaining attitude alone, but if one supply system fails, a propellant crossfeed allows one system to supply all 16 engines. Additionally, interconnect valves permit the RCS system to draw from ascent engine propellant tanks.

-more-

The engine clusters are mounted on outriggers 90 degree apart on the ascent stage.

The RCS provides small stabilizing impulses during ascent and descent burns, controls LM attitude during maneuvers, and produces thrust for separation and ascent/descent engine tank ullage. The system may be operated in either pulsed or steady-state modes.

Descent Propulsion System -- Maximum rated thrust of the descent engine is 9,870 pounds (4,380.9 kg) and is throttleable between 1,050 pounds (476.7 kg) and 6,300 pounds (2,860.2 kg). The engine can be gimballed six degrees in any direction for offset center of gravity trimming. Propellants are helium-pressurized Aerozine 50 and nitrogen tetroxide.

Ascent Propulsion System -- The 3,500 pound (1,589 kg) thrust ascent engine is not gimballed and performs at full thrust. The engine remains dormant until after the ascent stage separates from the descent stage. Propellants are the same as are burned in the RCS engines and the descent engine.

Caution and Warning, Controls and Displays -- These two systems have the same function aboard the lunar module as they do aboard the command module. (See CSM systems section.)

Tracking and Docking Lights -- A flashing tracking light (once per second, 20 milliseconds duration) on the front face of the lunar module is an aid for contingency CSM-active rendezvous LM rescue. Visibility ranges from 400 nautical miles through the CSM sextant to 130 miles with the naked eye. Five docking lights analagous to aircraft running lights are mounted on the LM for CSM-active rendezvous: two forward yellow lights, aft white light, port red light and starboard green light. All docking lights have about a 1,000-foot visibility.

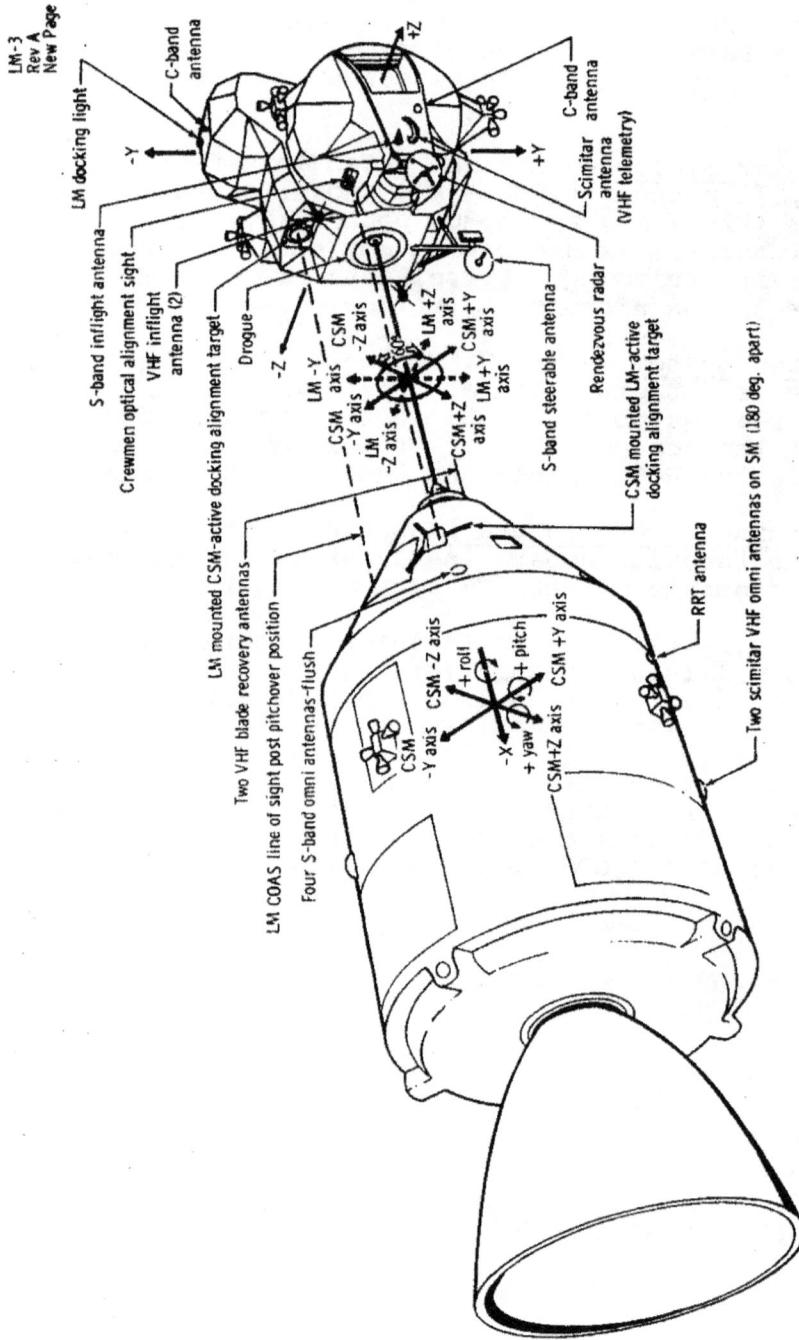

Figure 8-3.2. – LM-CSM antenna locations.

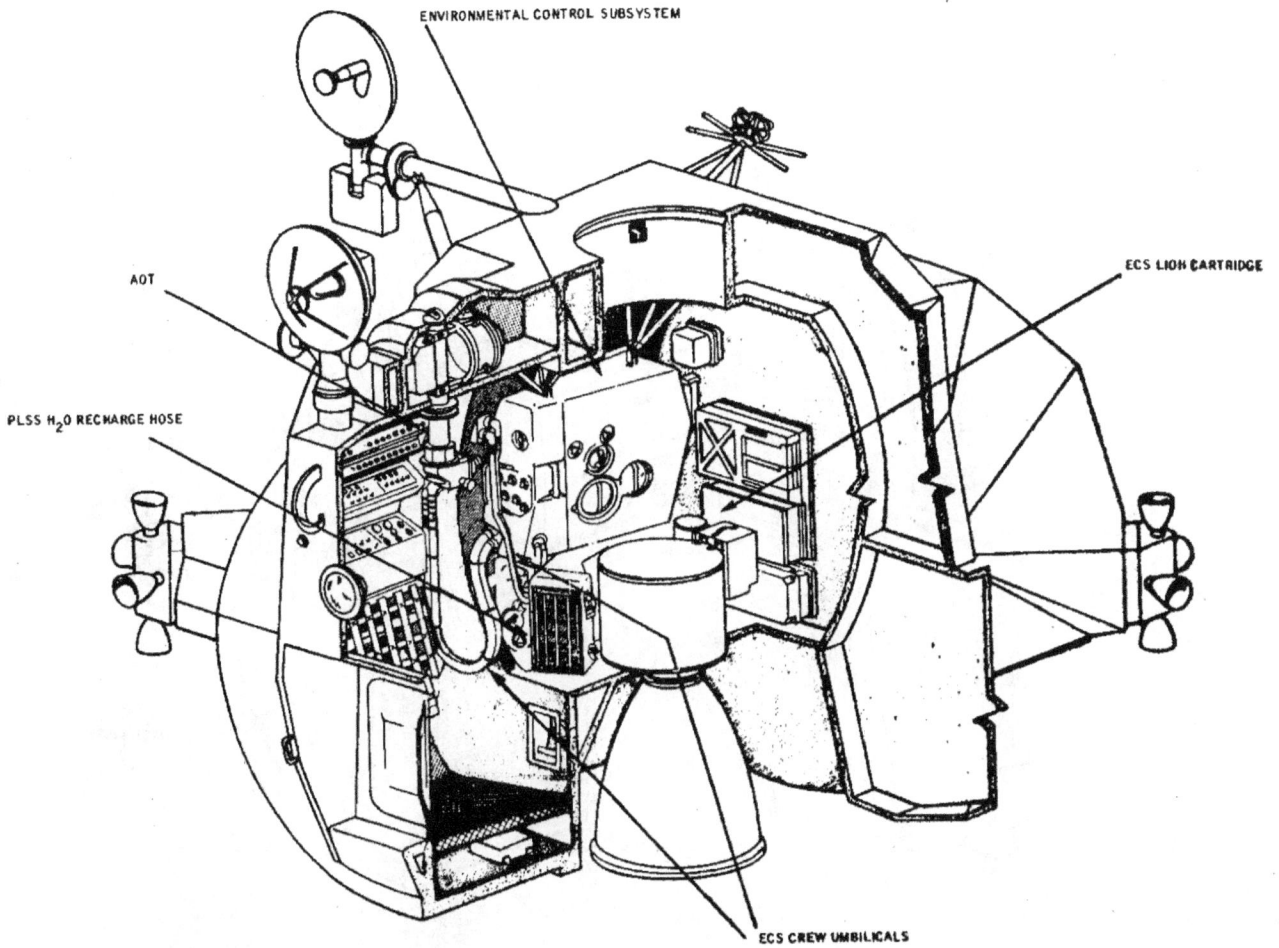

ENVIRONMENTAL CONTROL SUBSYSTEM

AOT

ECS LIOH CARTRIDGE

PLSS H$_2$0 RECHARGE HOSE

ECS CREW UMBILICALS

Cutaway of LM cabin interior, left half

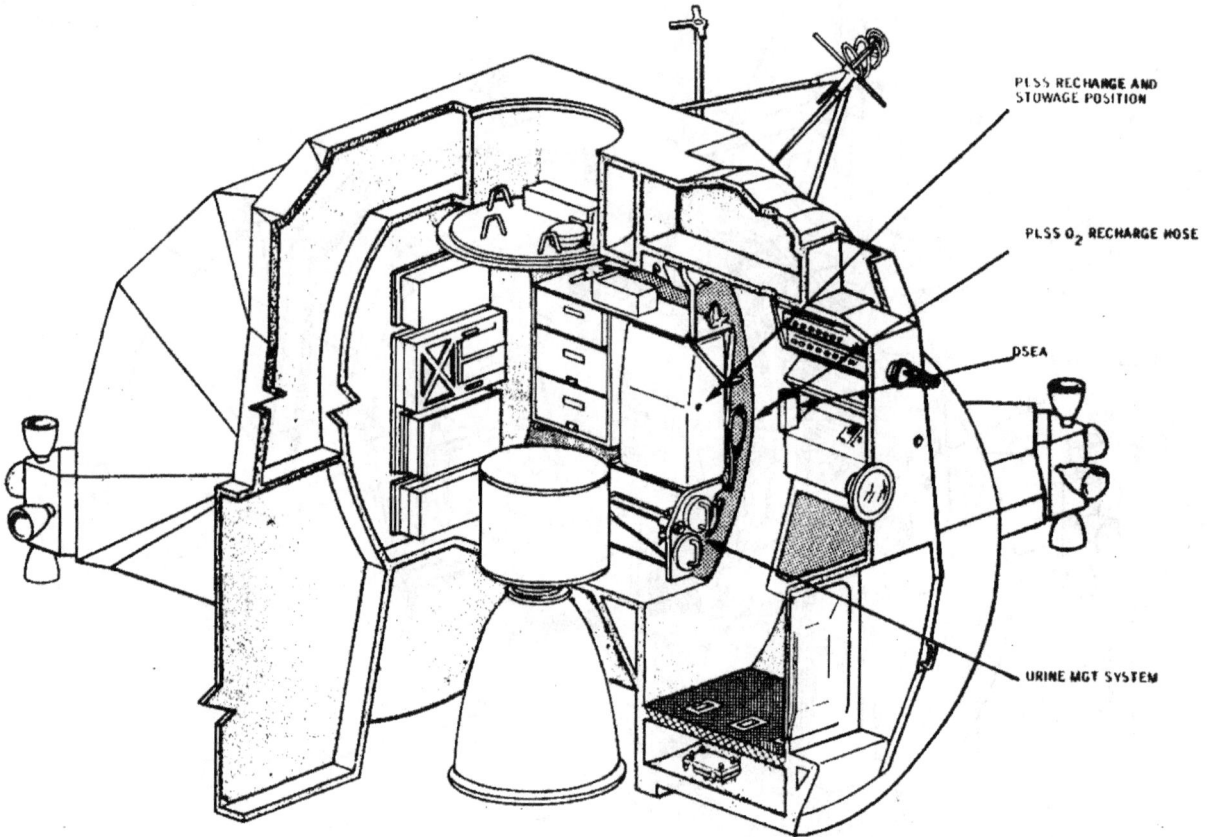

PLSS RECHARGE AND
STOWAGE POSITION

PLSS O_2 RECHARGE HOSE

DSEA

URINE MGT SYSTEM

Cutaway of LM cabin interior, right half

FACT SHEET, SA-504

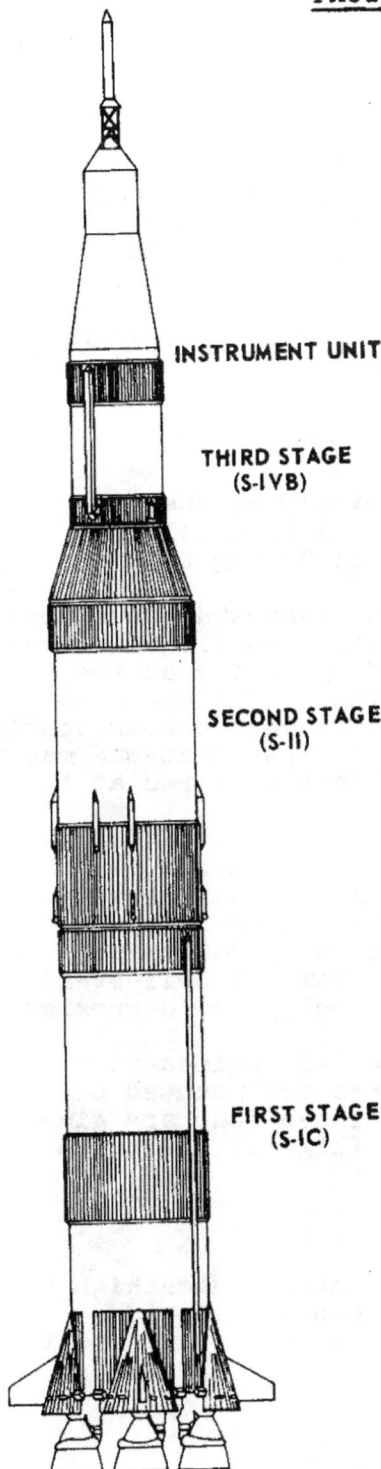

First Stage (S-IC)
Diameter-------- 33 feet, Height -- 138 feet
Weight---------- 5,026,200 lbs. fueled
                 295,600 lbs. dry
Engines-------- Five F-1
Propellants---- Liquid oxygen (347,300 gals.)
                RP-1 (Kerosene) - (211,140
                  gals.)
Thrust--------- 7,700,000 lbs.

Second Stage (S-II)
Diameter-------- 33 feet, Height -- 81.5 feet
Weight---------- 1,069,033 lbs. fueled
                 84,600 lbs. dry
Engines-------- Five J-2
Propellants---- Liquid oxygen (86,700 gals.)
                Liquid hydrogen (281,550
                  gals.)
Thrust--------- 1,150,000 lbs.
Interstage----- 10,305 lbs.

Third Stage (S-IVB)
Diameter-------- 21.7 feet, Height -- 58.3 ft.
Weight---------- 258,038 lbs. fueled
                 25,300 lbs. dry
Engines-------- One J-2
Propellants---- Liquid oxygen (19,600 gals)
                Liquid hydrogen (77,675
                  gals.)
Thrust--------- 232,000 lbs. (first burn)
                211,000 lbs. (second burn)
Interstage----- 8,081 lbs.

Instrument Unit
Diameter-------- 21.7 feet, Height -- 3 feet
Weight---------- 4,295 lbs.

NOTE: Weights and measures given above
are for the nominal vehicle configuration.
The figures may vary slightly due to changes
before or during flight to meet changing
conditions.

INSTRUMENT UNIT

THIRD STAGE
(S-IVB)

SECOND STAGE
(S-II)

FIRST STAGE
(S-IC)

LAUNCH VEHICLE

Saturn V

The Saturn V, 363 feet tall with the Apollo spacecraft in place, generates enough thrust to place a 125-ton payload into a 105 nm circular Earth orbit or boost a smaller payload to the vicinity of any planet in the solar system. It can boost about 50 tons to lunar orbit. The thrust of the three propulsive stages range from more than 7.7 million pounds for the booster to 230,000 pounds for the third stage at operating altitude. Including the instrument unit, the launch vehicle is 281 feet tall.

First Stage

The first stage (S-IC) was developed jointly by the National Aeronautics and Space Administration's Marshall Space Flight Center, Huntsville, Ala., and the Boeing Co.

The Marshall Center assembled four S-IC stages: a structural test model, a static test version and the first two flight stages. Subsequent flight stages are being assembled by Boeing at the Michoud Assembly Facility in New Orleans. The S-IC stage destined for the Apollo 9 mission was the first flight booster static tested at the NASA-Mississippi Test Facility. That test was made on May 11, 1967. Earlier flight stages were static fired at the NASA-Marshall Center.

The S-IC stage provides first boost of the Saturn V launch vehicle to an altitude of about 37 nautical miles (41.7 statute miles, 67.1 kilometers) and provides acceleration to increase the vehicle's velocity to 9,095 feet per second (2,402 m/sec, 5,385 knots, 6,201 mph). It then separates from the S-II stage and falls to Earth about 361.9 nm (416.9 sm, 667.3 km) downrange.

Normal propellant flow rate to the five F-1 engines is 29,522 pounds per second. Four of the engines are mounted on a ring, each 90 degrees from its neighbor. These four are gimballed to control the rocket's direction of flight. The fifth engine is mounted rigidly in the center.

Second Stage

The second stage (S-II), like the third stage, uses high performance J-2 engines that burn liquid oxygen and liquid hydrogen. The stage's purpose is to provide second stage boost nearly to Earth orbit.

At engine cutoff, the S-II separates from the third stage and, following a ballistic trajectory, plunges into the Atlantic Ocean about 2,412 nm (2,778.6 sm, 4,468 km) downrange from Cape Kennedy.

Five J-2 engines power the S-II. The four outer engines are equally spaced on a 17.5-foot diameter circle. These four engines may be gimballed through a plus or minus seven-degree square pattern for thrust vector control. Like the first stage, the center (number 5) engine is mounted on the stage centerline and is fixed.

The S-II carries the rocket to an altitude of 103 nm (118.7 sm, 190.9 km) and a distance of some 835 nm (961.9 sm, 1,548 km) downrange. Before burnout, the vehicle will be moving at a speed of 23,000 fps or 13,642 knots (15,708 mph, 25,291 kph, 6,619 m/sec). The J-2 engines will burn six minutes 11 seconds during this powered phase.

The Space Division of North American Rockwell Corp. builds the S-II at Seal Beach, Calif. The cylindrical vehicle is made up of the forward skirt (to which the third stage attaches), the liquid hydrogen tank, the liquid oxygen tank, the thrust structure (on which the engines are mounted) and an interstage section (to which the first stage connects). The propellant tanks are separated by an insulated common bulkhead.

The S-II was static tested by North American Rockwell at the NASA-Mississippi Test Facility on Feb. 10, 1968. This Apollo 9 flight stage was shipped to the test site via the Panama Canal for the test firing.

Third Stage

The third stage (S-IVB) was developed by the McDonnell Douglas Astronautics Co. at Huntington Beach, Calif. At Sacramento, Calif., the stage passed a static firing test on Aug. 26, 1968, as part of the preparation for the Apollo 9 mission. The stage was flown directly to the NASA-Kennedy Space Center.

Measuring 58 feet 4 inches long and 21 feet 8 inches in diameter, the S-IVB weighs 25,300 pounds dry. At first ignition it weighs 259,377 pounds. The interstage section weighs an additional 8,081 pounds. The stage's J-2 engine burns liquid oxygen and liquid hydrogen.

The stage, with its single engine, provides propulsion three times during the Apollo 9 mission. The first burn occurs immediately after separation from the S-II. It will last long enough (112 seconds) to insert the vehicle and spacecraft into Earth parking orbit. The second burn, which begins after separation from the spacecraft, will place the stage and instrument unit into a high apogee elliptical orbit. The third burn will drive the stage into solar orbit.

The fuel tanks contain 77,675 gallons of liquid hydrogen and 19,600 gallons of liquid oxygen at first ignition, totalling 230,790 pounds of propellants. Insulation between the two tanks is necessary because the liquid oxygen, at about 293 degrees below zero F., is warm enough, relatively, to heat the liquid hydrogen, at 423 degrees below zer F., rapidly and cause it to change into a gas.

The first reignition burn is for 62 seconds and the second reignition burn is planned to last 4 minutes 2 seconds. Both reignitions will be inhibited with inhibit removal to be by ground command only after separation of the spacecraft to a safe distance.

Instrument Unit

The Instrument Unit (IU) is a cylinder three feet high and 21 feet 8 inches in diameter. It weighs 4,295 pounds and contains the guidance, navigation and control equipment which will steer the vehicle through its Earth orbits and into the final escape orbit maneuver.

The IU also contains telemetry, communications, tracking and crew safety systems, along with its own supporting electrical power and environmental control systems.

Components making up the "brain" of the Saturn V are mounted on cooling panels fastened to the inside surface of the instrument unit skin. The "cold plates" are part of a system that removes heat by circulating cooled fluid through a heat exchanger that evaporates water from a separate supply into the vacuum of space.

The six major systems of the instrument unit are structural, thermal control, guidance and control, measuring and telemetry, radio frequency and electrical.

The instrument unit provides navigation, guidance and control of the vehicle; measurement of vehicle performance and environment; data transmission with ground stations; radio tracking of the vehicle; checkout and monitoring of vehicle functions; initiation of stage functional sequencing; detection of emergency situations; generation and network distribution of electric power for system operation; and preflight checkout and launch and flight operations.

A path-adaptive guidance scheme is used in the Saturn V instrument unit. A programmed trajectory is used in the initial launch phase with guidance beginning only after the vehicle has left the atmosphere. This is to prevent movements that might cause the vehicle to break apart while attempting to compensate for winds, jet streams and gusts encountered in the atmosphere.

If such air currents displace the vehicle from the optimum trajectory in climb, the vehicle derives a new trajectory. Calculations are made about once each second throughout the flight. The launch vehicle digital computer and launch vehicle data adapter perform the navigation and guidance computations.

The ST-124M inertial platform -- the heart of the navigation, guidance and control system -- provides space-fixed reference coordinates and measures accleration along the three mutually perpendicular axes of the coordinate system.

International Business Machines Corp., is prime contractor for the instrument unit and is the supplier of the guidance signal processor and guidance computer. Major suppliers of instrument unit components are: Electronic Communications, Inc., control computer; Bendix Corp., ST-124M inertial platform; and IBM Federal Systems Division, launch vehicle digital computer and launch vehicle data adapter.

Propulsion

The 41 rocket engines of the Saturn V have thrust ratings ranging from 72 pounds to more than 1.5 million pounds. Some engines burn liquid propellants, others use solids.

The five F-1 engines in the first stage burn RP-1 (kerosene) and liquid oxygen. Each engine in the first stage develops an average of 1,544,000 pounds of thrust at liftoff, building up to an average of 1,833,900 pounds before cutoff. The cluster of five engines gives the first stage a thrust range from 7.72 million pounds at liftoff to 9,169,560 pounds just before center engine cutoff.

The F-1 engine weighs almost 10 tons, is more than 18 feet high and has a nozzle-exit diameter of nearly 14 feet. The F-1 undergoes static testing for an average 650 seconds in qualifying for the 150-second run during the Saturn V first stage booster phase. This run period, 800 seconds, is still far less than the 2,200 seconds of the engine guarantee period. The engine consumes almost three tons of propellants per second.

The first stage of the Saturn V for this mission has eight other rocket motors. These are the solid-fuel retro-rockets which will slow and separate the stage from the second stage. Each rocket produces a thrust of 87,900 pounds for 0.6 second.

The main propulsion for the second stage is a cluster of
five J-2 engines burning liquid hydrogen and liquid oxygen.
Each engine develops a mean thrust of more than 205,000 pounds
at 5.0:1 mixture ratio (variable from 193,000 to 230,000 in
phases of flight), giving the stage a total mean thrust of more
than a million pounds.

Designed to operate in the hard vacuum of space, the 3,500-
pound J-2 is more efficient than the F-1 because it burns the
high-energy fuel hydrogen.

The second stage also has four 21,000-pound-thrust solid
fuel rocket engines. These are the ullage rockets mounted on
the S-IC/S-II interstage section. These rockets fire to settle
liquid propellant in the bottom of the main tanks and help at-
tain a "clean" separation from the first stage, they remain with
the interstage when it drops away at second plane separation.
Four retrorockets are located in the S-IVB aft interstage (which
never separates from the S-II) to separate the S-II from the
S-IVB prior to S-IVB ignition.

Eleven rocket engines perform various functions on the
third stage. A single J-2 provides the main propulsive force;
there are two jettisonable main ullage rockets and eight smaller
engines in the two auxiliary propulsion system modules.

Launch Vehicle Instrumentation and Communication

A total of 2,159 measurements will be taken in flight on
the Saturn V launch vehicle: 666 on the first stage, 975 on
the second stage, 296 on the third stage and 222 on the instrument unit.

The Saturn V will have 16 telemetry systems: six on the
first stage, six on the second stage, one on the third stage
and three on the instrument unit. A radar tracking system will
be on the first stage, and a C-Band system and command system
on the instrument unit. Each powered stage will have a range
safety system as on previous flights.

There will be no film or television cameras on or in any
of the stages of the Saturn 504 launch vehicle.

-more-

Vehicle Weights During Flight

Event	Vehicle Weight
Ignition	6,486,915 pounds
Liftoff	6,400,648 pounds
Mach 1	4,487,938 pounds
Max. Q	4,015,350 pounds
CECO	2,443,281 pounds
OECO	1,831,574 pounds
S-IC/S-II Separation	1,452,887 pounds
S-II Ignition	1,452,277 pounds
Interstage Jettison	1,379,282 pounds
LET Jettison	1,354,780 pounds
S-II Cutoff	461,636 pounds
S-II/S-IVB Separation	357,177 pounds
S-IVB Ignition	357,086 pounds
S-IVB First Cutoff	297,166 pounds
Parking Orbit Injection	297,009 pounds
Spacecraft First Separation	232,731 pounds (Only S-IVB and LM)
Spacecraft Docking	291,572 pounds (S-IVB and complete spacecraft)
Spacecraft Second Separation	199,725 pounds (only S-IVB stage)
S-IVB First Reignition	199,346 pounds

Event	Vehicle Weight
S-IVB Second Cutoff	170,344 pounds
Intermediate Orbit Injection	170,197 pounds
S-IVB Second Reignition	169,383 pounds
S-IVB Third Cutoff	54,440 pounds
Escape Orbit Injection	54,300 pounds
End LOX Dump	35,231 pounds
End LH2 Dump	31,400 pounds

S-IVB Restarts

The third stage (S-IVB) of the Saturn V rocket for the Apollo 9 mission will burn a total of three times in space, the last two burns unmanned for engineering evaluation of stage capability. It has never burned more than twice in space before.

Engineers also want to check out the primary and backup propellant tank pressurization systems and prove that in case the primary system fails the backup system will pressurize the tanks sufficiently for restart.

Also planned for the second restart is an extended fuel lead for chilldown of the J-2 engine using liquid hydrogen fuel flowing through the engine to chill it to the desired temperature prior to restart.

All these events -- second restart, checkout of the backup pressurization system and extended fuel lead chilldown will not affect the primary mission of Apollo 9. The spacecraft will have been separated and will be safe in a different orbit from that of the spacecraft before the events begin.

Previous flights of the S-IVB stage have not required a second restart, and none of the flights currently planned have a specific need for the third burn.

The second restart during the Apollo 9 mission will come 80 minutes after second engine burn cutoff to demonstrate a requirement that the stage has the capability to restart in space after being shut down for only 80 minutes. That capability has not yet been proven because flights to date have not required as little as 80 minutes of coastings.

-more-

The first restart is scheduled for four hours 45 minutes and 41 seconds after launch, or about four and one-half hours after first burn cutoff. The need for a second restart may occur on future non-lunar flights. Also, the need to launch on certain days could create a need for the 80-minute restart capability.

The primary pressurization system of the propellant tanks for S-IVB restart uses a helium heater. In this system, nine helium storage spheres in the liquid hydrogen tank contain gaseous helium charged to about 3,000 psi. This helium is passed through the heater which heats and expands the gas before it enters the propellant tanks. The heater operates on hydrogen and oxygen gas from the main propellant tanks.

The backup system consists of five ambient helium spheres mounted on the stage thrust structure. This system, controlled by the fuel repressurization control module, can repressurize the tanks in case the primary system fails.

The first restart will use the primary system. If that system fails, the backup system will be used. The backup system will be used for the second restart.

The primary reason for the extended fuel lead chilldown test is to demonstrate a contingency plan in case the chilldown pumps fail.

In the extended chilldown event, a ground command will cut off the pumps to put the stage in a simulated failure condition. Another ground command will start liquid hydrogen flowing through the engine.

This is a slower method, but enough time has been allotted for the hydrogen to chill the engine and create about the same conditions as would be created by the primary system.

The two unmanned restarts in this mission are at the very limits of the design requirements for the S-IVB and the J-2 engine. For this reason, the probability of restart is not as great as in the nominal lunar missions.

Two requirements are for the engine to restart four and one-half hours after first burn cutoff, and for a total stage lifetime of six and one-half hours. The first restart on this mission will be four and one-half hours after first burn cutoff, and second restart will occur six hours and six minutes after liftoff, both events scheduled near the design limits.

Differences in Apollo 8 and Apollo 9 Launch Vehicles

Two modifications resulting from problems encountered during the second Saturn V flight were incorporated and proven successful on the third Saturn V mission. The new helium pre-valve cavity pressurization system will again be flown on the S-IC stage of Apollo 9. Also, new augmented spark igniter lines which flew on the engines of the two upper stages of Apollo 8 will again be used on Apollo 9.

The major S-IC stage differences between Apollo 8 and Apollo 9 are:

1. Dry weight was reduced from 304,000 pounds to 295,600 pounds.

2. Weight at ground ignition increased from 4,800,000 pounds to 5,026,200 pounds.

3. Instrumentation measurements were reduced from 891 to 666.

4. Camera instrumentation electrical power system is not installed on S-IC-4.

5. S-IC-4 carries neither a TV camera system nor a film camera system.

The Saturn V will fly a somewhat lighter and slightly more powerful second stage beginning with Apollo 9.

The changes are:

1. Nominal vacuum thrust for J-2 engines was increased from 225,000 pounds each to 230,000 pounds each. This changed the second stage thrust from a total of 1,125,000 pounds to 1,150,000 pounds.

2. The approximate empty S-II stage weight has been reduced from 88,000 to 84,600 pounds. The S-IC/S-II interstage weight was reduced from 11,800 to 11,664 pounds.

3. Approximate stage gross liftoff weight was increased from 1,035,000 pounds to 1,069,114 pounds.

4. S-II instrumentation system was changed from research and development to a combination of research and development and operational.

-more-

5. Instrumentation measurements were decreased from 1,226 to 975.

Major differences between the S-IVB stage used on Apollo 8 and the one on Apollo 9 are:

1. S-IVB dry stage weight decreased from 26,421 pounds to 25,300 pounds. This does not include the 8,081-pound interstage section.

2. S-IVB gross stage weight at liftoff decreased from 263,204 pounds to 259,337 pounds.

3. Stage measurements evolved from research and development to operational status.

4. Instrumentation measurements were reduced from 342 to 296.

Major instrument unit differences between Apollo 8 and Apollo 9 include deletions of a rate gyro timer, thermal probe, a measuring distributor, a tape recorder, two radio frequency assemblies, a source follower, a battery and six measuring racks. Instrumentation measurements were reduced from 339 to 222.

Launch Vehicle Sequence of Events

(Note: Information presented in this press kit is based upon a nominal mission. Plans may be altered prior to or during flight to meet changing conditions.)

Launch

The first stage of the Saturn V will carry the launch vehicle and Apollo spacecraft to an altitude of 36.2 nautical miles (41.7 sm, 67.1 km) and 50 nautical miles (57.8 sm, 93 km) downrange, building up speed to 9,095.2 feet per second (2,402 m/sec, 5,385.3 knots, 6,201.1 sm) in two minutes 31 seconds of powered flight. After separation from the second stage, the first stage will continue on a ballistic trajectory ending in the Atlantic Ocean some 361.9 nautical miles (416.9 sm, 670.9 km) downrange from Cape Kennedy (latitude 30.27 degrees north and longitude 73.9 degrees west) about nine minutes after liftoff.

Second Stage

The second stage, with engines running six minutes and 11 seconds, will propel the vehicle to an altitude of about 103 nautical miles (118.7 sm, 190.9 km) some 835 nautical miles (961.7 sm, 1,547 km) downrange, building up to 23,040.3 feet per second (6,619 m/sec, 13,642.1 knots, 15,708.9 mph) space fixed velocity. The spent second stage will land in the Atlantic Ocean about 20 minutes after liftoff some 2,410.3 nautical miles (2,776.7 sm, 4,468.4 km) from the launch site, at latitude 31.46 degrees north and longitude 34.06 degrees west.

Third Stage First Burn

The third stage, in its 112-second initial burn, will place itself and the Apollo spacecraft into a circular orbit 103 nautical miles (119 sm, 191 km) above the Earth. Its inclination will be 32.5 degrees and the orbital period about 88 minutes. Apollo 9 will enter orbit at about 56.66 degrees west longitude and 32.57 degrees north latitude.

Parking Orbit

The Saturn V third stage will be checked out in Earth parking orbit in preparation for the second S-IVB burn. During the second revolution, the Command/Service Module (CSM) will separate from the third stage. The spacecraft LM adapter (SLA) panels will be jettisoned and the CSM will turn around and dock with the lunar module (LM) while the LM is still attached to the S-IVB/IU. This maneuver is scheduled to require 14 minutes. About an hour and a quarter after the docking, LM ejection is to occur. A three-second SM RCS burn will provide a safe separation distance at S-IVB reignition. All S-IVB reignitions are nominally inhibited. The inhibit is removed by ground command after the CSM/LM is determined to be a safe distance away.

Third Stage Second Burn

Boost of the unmanned S-IVB stage from Earth parking orbit to an intermediate orbit occurs during the third revolution shortly after the stage comes within range of Cape Kennedy (4 hours 44 minutes 42 seconds after liftoff). The second burn, lasting 62 seconds, will put the S-IVB into an elliptical orbit with an apogee of 3,052 km (1,646.3 nm, 1,896.5 sm) and a perigee of 196 km (105.7 nm, 121.8 sm). The stage will remain in this orbit about one-half revolution.

-more-

Third Stage Third Burn

The third stage is reignited at 6 hours 6 minutes 4 seconds after liftoff for a burn lasting 4 minutes 2 seconds. This will place the stage and instrument unit into the escape orbit. Ninety seconds after cutoff, LOX dump begins and lasts for 11 minutes 10 seconds, followed 10 seconds later by the dump of the liquid hydrogen, an exercise of 18 minutes 15 seconds duration. Total weight of the stage and instrument unit placed into solar orbit will be about 31,400 pounds.

Launch Vehicle Key Events

Time Hrs	Min	Sec	Event	Altitude Meters	Feet	N. Mi.	S. Mi.	Velocity M/Sec.	F/Sec.	MPH	Knots
00	00	00	First Motion	60	196	.033	.038	0.00	0.00	0.0	0.00
00	00	12	Tilt Initiation	225	737	.12	.14	410.00	1345.4	279.5	242.8
00	01	21	Maximum Dynamic Pressure	13311	43671	7.2	8.3	816.00	2677.00	1825.2	1585.1
00	02	14	Center Engine Cutoff	44981	147576	24.3	27.9	2005.6	6580.2	4486.4	3896.1
00	02	39	Outboard Engine Cutoff	67132	220248	36.2	41.7	2772.2	9095.2	6201.1	5385.3
00	02	40	S-IC/S-II Separation	67848	222599	36.6	42.2	2781.9	9127.	6222.8	5404.1
00	02	42	S-II Ignition	69354	227541	37.4	43.1	2777.7	9113.3	6213.3	5396.0
00	03	10	S-II Aft Interstage Jetn.	92810	304514	50.1	57.7	2909.6	9546.	6508.5	5652.2
00	03	15	LES Jettison	97075	318487	52.4	60.2	2942.6	9654.2	6582.2	5716.3
00	03	21	Initiate IGM	100747	330534	54.3	62.0	2973.3	9754.8	6650.8	5775.8
00	08	53	S-II Cutoff	190957	626497	103.0	118.7	7022.7	23040.3	15708.9	13642.2
00	08	54	S-II/S-IVB Separation	191045	626653	103.0	118.7	7027.3	23055.5	15719.2	13651.2
00	08	57	S-IVB Ignition	191137	627089	103.1	118.6	7027.6	23056.5	15719.2	13651.6
00	10	49	S-IVB First Cutoff	191385	627904	103.2	118.9	7791.	25561.1	17427.6	15134.7
00	10	59	Parking Orbit Insertion	191398	627947	103.2	118.9	7793.	25567.7	17432.1	15138.6
02	43	43	Spacecraft Separation	195960	642923	105.7	121.6	7790.8	25560.4	17427.1	15134.3
02	53	57	Spacecraft Docking	198552	651416	107.1	123.4	7787.9	25551.	17420.7	15128.7
04	08	41	Spacecraft Final Sep'n.	194366	637682	104.8	120.8	7792.3	25556.2	17430.4	15137.2
04	45	43	S-IVB Reignition	199949	655671	107.8	124.2	7789.6	25556.6	17424.4	15131.9
04	46	43	S-IVB Second Cutoff	200175	656744	108.0	124.4	8449.	27719.8	18899.4	16412.9
06	06	53	Intermediate Orbit Inser.	200720	658532	108.3	124.7	8451.5	27728.2	18905.1	16417.9
06	07	04	S-IVB Reignition	2410165	7907364	1300.0	1497.6	6400.1	20997.8	14316.3	12432.8
06	11	05	S-IVB Third Cutoff	2242364	7356839	1209.5	1393.3	11230.8	36846.4	25121.9	21816.8
06	11	15	Escape Orbit Insertion	2244556	7364028	1210.7	1394.7	11239.3	36874.2	25146.8	21833.2
06	12	36	Start LOX Dump	2295412	7530878	1238.1	1426.3	11215.2	36795.3	25087.0	21786.5
06	23	46	LOX Dump Cutoff	4662286	15296215	2514.8	2897.0	10372.6	34030.8	23202.2	20149.6
06	23	56	Start LH2 Dump	4716809	15475094	2544.2	2930.9	10355.5	33974.6	23163.9	20116.4
06	42	11	LH2 Dump Cutoff	11929321	39138193	6434.5	7412.5	8911.5	29237.2	19933.9	17311.3

LAUNCH FACILITIES

Kennedy Space Center-Launch Complex

NASA's John F. Kennedy Space Center performs preflight checkout, test and launch of the Apollo 9 space vehicle. A government-industry team of about 550 will conduct the final countdown from Firing Room 2 of the Launch Control Center (LCC).

The firing room team is backed up by more than 5,000 persons who are directly involved in launch operations at KSC -- from the time the vehicle and spacecraft stages arrive at the center until the launch is completed.

Initial checkout of the Apollo spacecraft is conducted in work stands and in the altitude chambers in the Manned Space-craft Operations Building (MSOB) at Kennedy Space Center. After completion of checkout there, the assembled spacecraft is taken to the Vehicle Assembly Building (VAB) and mated with the launch vehicle. There the first integrated spacecraft and launch vehicle tests are conducted. The assembled space vehicle is then rolled out to the launch pad for final preparations and count-down to launch.

In August 1968 a decision was made not to fly Lunar Module 3 on Apollo 8 as had been originally planned. Checkout con-tinued with the remainder of the Apollo 8 space vehicle and Lunar Module 3 was integrated with the test schedule for Apollo 9.

LM-3 arrived at KSC in June 1968 and at the time the decision was made to fly it on Apollo 9 it had just completed systems tests in the MSOB. It was moved to the vacuum chamber later in August and four manned altitude chamber tests were conducted in September. During these tests, the chamber was pumped down to simulate altitudes in excess of 200,000 feet (60,960 meters) and the lunar module and crew systems were thoroughly checked. The prime crew of Spacecraft Commander James McDivitt and Lunar Module Pilot Russell Schweickart participated in two of the runs and the backup crew of Charles Conrad and Alan Bean participated in the other two runs.

The Apollo 9 command/service module arrived at KSC in October and after receiving inspection in the MSOB, a docking test was conducted with the LM. Two manned altitude chamber runs were made in November with the prime crew participating in one and the backups in the other.

In December the LM was mated to the spacecraft lunar module adapter (SLA), the command/service module was mated to the SLA, and the assembled spacecraft was moved to the VAB where it was erected on the Saturn V launch vehicle (SA 504).

The Apollo 9 launch vehicle had been assembled on its mobile launcher in the VAB in early October. Tests were conducted on individual systems on each of the stages and on the overall vehicle before the spacecraft was mated.

After spacecraft erection, the spacecraft and launch vehicle were electrically mated and the first overall test (plugs-in) of the space vehicle was conducted. In accordance with the philosophy of accomplishing as much of the checkout as possible in the VAB, the overall test was conducted before the space vehicle was moved to the launch pad.

The plugs-in test verified the compatibility of the space vehicle systems, ground support equipment and off-site support facilities by demonstrating the ability of the systems to proceed through a simulated countdown, launch and flight. During the simulated flight portion of the test, the systems were required to respond to both emergency and normal flight conditions.

The move to the launch pad was conducted Jan. 3. Because minimum pad damage was incurred from the launch of Apollo 8, it was possible to refurbish the pad and roll out Apollo 9 less than two weeks later. The 3½-mile (5.6 km) trip to the pad aboard the transporter was completed in about eight hours.

The space vehicle Flight Readiness Test was conducted late in January. Both the prime and backup crews participate in portions of the FRT, which is a final overall test of the space vehicle systems and ground support equipment when all systems are as near as possible to a launch configuration.

After hypergolic fuels were loaded aboard the space vehicle, the launch vehicle first stage fuel (RP-1) was brought aboard and the final major test of the space vehicle began. This was the countdown demonstration test (CDDT), a dress rehearsal for the final countdown to launch. The CDDT for Apollo 9 was divided into a "wet" and a "dry" portion. During the first, or "wet" portion, the entire countdown, including propellant loading, was carried out down to T-8.9 seconds. The astronaut crews did not participate in the wet CDDT. At the completion of the wet CDDT, the cryogenic propellants (liquid oxygen and liquid hydrogen) were off-loaded, and the final portion of the countdown was re-run, this time simulating the fueling and with the prime astronaut crew participating as they will on launch day.

-more-

During the assembly and checkout operations for Apollo 9, launch crews at Kennedy Space Center completed the preparation and launch of Apollo 7 (Oct. 11, 1968) and Apollo 8 (Dec. 21, 1968) and began the assembly and checkout operations for Apollo 10 and Apollo 11 to be launched later this year.

Because of the complexity involved in the checkout of the 363-foot-tall (110.6 meters) Apollo/Saturn V configuration, the launch teams make use of extensive automation in their checkout. Automation is one of the major differences in checkout used on Apollo compared to the procedures used in the Mercury and Gemini programs.

RCA 110A computers, data display equipment and digital data techniques are used throughout the automatic checkout from the time the launch vehicle is erected in the VAB through liftoff. A similar, but separate computer operation called ACE (Acceptance Checkout Equipment) is used to verify the flight readiness of the spacecraft. Spacecraft checkout is controlled from separate firing rooms located in the Manned Spacecraft Operations Building.

KSC Launch Complex 39

Launch Complex 39 facilities at the Kennedy Space Center were planned and built specifically for the Saturn V program, the space vehicle that will be used to carry astronauts to the Moon.

Complex 39 introduced the mobile concept of launch operations, a departure from the fixed launch pad techniques used previously at Cape Kennedy and other launch sites. Since the early 1950's when the first ballistic missiles were launched, the fixed launch concept had been used on NASA missions. This method called for assembly, checkout and launch of a rocket at one site--the launch pad. In addition to tying up the pad, this method also often left the flight equipment exposed to the outside influences of the weather for extended periods.

Using the mobile concept, the space vehicle is thoroughly checked in an enclosed building before it is moved to the launch pad for final preparations. This affords greater protection, a more systematic checkout process using computer techniques, and a high launch rate for the future, since the pad time is minimal.

Saturn V stages are shipped to the Kennedy Space Center by ocean-going vessels and specially designed aircraft, such as the Guppy. Apollo spacecraft modules are transported by air. The spacecraft components are first taken to the Manned Spacecraft Operations Building for preliminary checkout. The Saturn V stages are brought immediately to the Vehicle Assembly Building after arrival at the nearby turning basin.

Apollo 9 is the fourth Saturn V to be launched from Pad A, Complex 39. The historic first launch of the Saturn V, designated Apollo 4, took place Nov. 9, 1967 after a perfect countdown and on-time liftoff at 7 a.m. EST. The second Saturn V mission--Apollo 6--was conducted last April 4. The third Saturn V mission, Apollo 8, was conducted last Dec. 21-27.

The major components of Complex 39 include: (1) the Vehicle Assembly Building (VAB) where the Apollo 9 was assembled and prepared; (2) the Launch Control Center, where the launch team conducts the preliminary checkout and count-down; (3) the mobile launcher, upon which the Apollo 9 was erected for checkout and from where it will be launched; (4) the mobile service structure, which provides external access to the space vehicle at the pad; (5) the transporter, which carries the space vehicle and mobile launcher, as well as the mobile service structure to the pad; (6) the crawlerway over which the space vehicle travels from the VAB to the launch pad; and (7) the launch pad itself.

The Vehicle Assembly Building

The Vehicle Assembly Building is the heart of Launch Complex 39. Covering eight acres, it is where the 363-foot-tall space vehicle is assembled and tested.

The VAB contains 129,482,000 cubic feet of space. It is 716 feet long, and 518 feet wide and it covers 343,500 square feet of floor space.

The foundation of the VAB rests on 4,225 steel pilings, each 16 inches in diameter, driven from 150 to 170 feet to bedrock. If placed end to end, these piles would extend a distance of 123 miles. The skeletal structure of the building contains approximately 60,000 tons of structural steel. The exterior is covered by more than a million square feet of insulated aluminum siding.

The building is divided into a high bay area 525 feet high and a low bay area 210 feet high, with both areas serviced by a transfer aisle for movement of vehicle stages.

The low bay work area, approximately 442 feet wide and 274 feet long, contains eight stage-preparation and check-out cells. These cells are equipped with systems to simulate stage interface and operation with other stages and the instrument unit of the Saturn V launch vehicle.

After the Apollo 9 launch vehicle upper stages arrived at the Kennedy Space Center, they were moved to the low bay of the VAB. Here, the second and third stages underwent acceptance and checkout testing prior to mating with the S-IC first stage atop mobile launcher No. 2 in the high bay area.

The high bay provides the facilities for assembly and checkout of both the launch vehicle and spacecraft. It contains four separate bays for vertical assembly and checkout. At present, three bays are equipped, and the fourth will be reserved for possible changes in vehicle configuration.

Work platforms -- some as high as three-story buildings -- in the high bays provide access by surrounding the launch vehicle at varying levels. Each high bay has five platforms. Each platform consists of two bi-parting sections that move in from opposite sides and mate, providing a 360-degree access to the section of the space vehicle being checked.

A 10,000-ton-capacity air conditioning system, sufficient to cool about 3,000 homes, helps to control the environment within the entire office, laboratory, and workshop complex located inside the low bay area of the VAB. Air conditioning is also fed to individual platform levels located around the vehicle.

There are 141 lifting devices in the VAB, ranging from one-ton hoists to two 250-ton high-lift bridge cranes.

The mobile launchers, carried by transporter vehicles, move in and out of the VAB through four doors in the high bay area, one in each of the bays. Each door is shaped like an inverted T. They are 152 feet wide and 114 feet high at the base, narrowing to 76 feet in width. Total door height is 456 feet.

The lower section of each door is of the aircraft hangar type that slides horizontally on tracks. Above this are seven telescoping vertical lift panels stacked one above the other, each 50 feet high and driven by an individual motor. Each panel slides over the next to create an opening large enough to permit passage of the Mobile Launcher.

The Launch Control Center

Adjacent to the VAB is the Launch Control Center (LCC). This four-story structure is a radical departure from the dome-shaped blockhouses at other launch sites.

The electronic "brain" of Launch Complex 39, the LCC was used for checkout and test operations while Apollo 9 was being assembled inside the VAB. The LCC contains display, monitoring, and control equipment used for both checkout and launch operations.

The building has telemeter checkout stations on its second floor, and four firing rooms, one for each high bay of the VAB, on its third floor. Three firing rooms will contain identical sets of control and monitoring equipment, so that launch of a vehicle and checkout of others may take place simultaneously. A ground computer facility is associated with each firing room.

The high speed computer data link is provided between the LCC and the mobile launcher for checkout of the launch vehicle. This link can be connected to the mobile launcher at either the VAB or at the pad.

The three equipped firing rooms have some 450 consoles which contain controls and displays required for the checkout process. The digital data links connecting with the high bay areas of the VAB and the launch pads carry vast amounts of data required during checkout and launch.

There are 15 display systems in each LCC firing room, with each system capable of providing digital information instantaneously.

Sixty television cameras are positioned around the Apollo/Saturn V transmitting pictures on 10 modulated channels. The LCC firing room also contains 112 operational intercommunication channels used by the crews in the checkout and launch countdown.

Mobile Launcher

The mobile launcher is a transportable launch base and umbilical tower for the space vehicle. Three launchers are used at Complex 39.

The launcher base is a two-story steel structure, 25 feet high, 160 feet long, and 135 feet wide. It is positioned on six steel pedestals 22 feet high when in the VAB or at the launch pad. At the launch pad, in addition to the six steel pedestals, four extendable colums also are used to stiffen the mobile launcher against rebound loads, if the engine cuts off.

The umbilical tower, extending 398 feet above the launch platform, is mounted on one end of the launcher base. A hammerhead crane at the top has a hook height of 376 feet above the deck with a traverse radius of 85 feet from the center of the tower.

The 12-million-pound mobile launcher stands 445 feet high when resting on its pedestals. The base, covering about half an acre, is a compartmented structure built of 25-foot steel girders.

The launch vehicle sits over a 45-foot-square opening which allows an outlet for engine exhausts into a trench containing a flame deflector. This opening is lined with a replaceable steel blast shield, independent of the structure, and will be cooled by a water curtain initiated two seconds after liftoff.

There are nine hydraulically-operated service arms on the umbilical tower. These service arms support lines for the vehicle umbilical systems and provide access for personnel to the stages as well as the astronaut crew to the spacecraft.

On Apollo 9, one of the service arms is retracted early in the count. The Apollo spacecraft access arm is partially retracted at T-43 minutes. A third service arm is released at T-30 seconds, and a fourth at about T-6 seconds. The remaining five arms are set to swing back at vehicle first motion after T-0.

The service arms are equipped with a backup retraction system in case the primary mode fails.

The Apollo access arm (service arm No. 9), located at the 320-foot level above the launcher base, provides access to the spacecraft cabin for the closeout team and astronaut crews. The flight crew will board the spacecraft starting at about T-2 hours, 40 minutes in the count. The access arm will be moved to a parked position, 12 degrees from the spacecraft, at about T-43 minutes.

This is a distance of about three feet, which permits a rapid reconnection of the arm to the spacecraft in the event of an emergency condition. The arm is fully retracted at the T-5 minute mark in the count.

The Apollo 9 vehicle is secured to the mobile launcher by four combination support and hold-down arms mounted on the launcher deck. The hold-down arms are cast in one piece, about 6 X 9 feet at the base and 10 feet tall, weighing more than 20 tons. Damper struts secure the vehicle near its top.

After the engines ignite, the arms hold Apollo 9 for about six seconds until the engines build up to 95 per cent thrust and other monitored systems indicate they are functioning properly. The arms release on receipt of a launch commit signal at the zero mark in the count. But the vehicle is prevented from accelerating too rapidly by controlled release mechanisms.

The Transporter

The six-million-pound transporters, the largest tracked vehicles known, move mobile launchers into the VAB and mobile launchers with assembled Apollo space vehicles to the launch pad. They also are used to transfer the mobile service structure to and from the launch pads. Two transporters are in use at Complex 39.

The Transporter is 131 feet long and 114 feet wide. The vehicle moves on four double-tracked crawlers, each 10 feet high and 40 feet long. Each shoe on the crawler tracks seven feet six inches in length and weighs about a ton.

Sixteen traction motors powered by four 1,000-kilowatt generators, which in turn are driven by two 2,750-horsepower diesel engines, provide the motive power for the transporter. Two 750-kw generators, driven by two 1,065-horsepower diesel engines, power the jacking, steering, lighting, ventilating and electronic systems.

Maximum speed of the transporter is about one-mile-per-hour loaded and about two-miles-per-hour unloaded. A 3½ mile trip to the pad with a mobile launcher, made at less than maximum speed, takes approximately seven hours.

The transporter has a leveling system designed to keep the top of the space vehicle vertical within plus-or-minus 10 minutes of arc -- about the dimensions of a basketball.

This system also provides leveling operations required to negotiate the five per cent ramp which leads to the launch pad, and keeps the load level when it is raised and lowered on pedestals both at the pad and within the VAB.

The overall height of the transporter is 20 feet from ground level to the top deck on which the mobile launcher is mated for transportation. The deck is flat and about the size of a baseball diamond (90 by 90 feet).

Two operator control cabs, one at each end of the chassis located diagonally opposite each other, provide totally en-closed stations from which all operating and control functions are coordinated.

The transporter moves on a roadway 131 feet wide, divided by a median strip. This is almost as broad as an eight-lane turnpike and is designed to accommodate a combined weight of about 18 million pounds.

The roadway is built in three layers with an average depth of seven feet. The roadway base layer is two-and-one-half feet of hydraulic fill compacted to 95 per cent density. The next layer consists of three feet of crushed rock packed to maximum density, followed by a layer of one foot of selected hydraulic fill. The bed is topped and sealed with an asphalt prime coat.

-more-

On top of the three layers is a cover of river rock,
eight inches deep on the curves and six inches deep on the
straightway. This layer reduces the friction during steering
and helps distribute the load on the transporter bearings.

Mobile Service Structure

A .402-foot-tall, 9.8-million-pound tower is used to
service the Apollo launch vehicle and spacecraft at the pad.
The 40-story steel-trussed tower, called a mobile service
structure, provides 360-degree platform access to the Saturn
vehicle and the Apollo spacecraft.

The service structure has five platforms -- two self-
propelled and three fixed, but movable. Two elevators carry
personnel and equipment between work platforms. The platforms
can open and close around the 363-foot space vehicle.

After depositing the mobile launcher with its space
vehicle on the pad, the transporter returns to a parking area
about 7,000 feet from the pad. There it picks up the mobile
service structure and moves it to the launch pad. At the
pad, the huge tower is lowered and secured to four mount
mechanisms.

The top three work platforms are located in fixed
positions which serve the Apollo spacecraft. The two lower
movable platforms serve the Saturn V.

The mobile service structure ramains in position until
about T-11 hours when it is removed from its mounts and returned
to the parking area.

Water Deluge System

A water deluge system will provide a million gallons
of industrial water for cooling and fire prevention during
launch of Apollo 9. Once the service arms are retracted at
liftoff, a spray system will come on to cool these arms from
the heat of the five Saturn F-1 engines during liftoff.

On the deck of the mobile launcher are 29 water nozzles.
This deck deluge will start immediately after liftoff and will
pour across the face of the launcher for 30 seconds at the rate
of 50,000 gallons-per-minute. After 30 seconds, the flow will
be reduced to 20,000 gallons-per-minute.

-more-

Positioned on both sides of the flame trench are a
series of nozzles which will begin pouring water at 8,000
gallons-per-minute, 10 seconds before liftoff. This water will
be directed over the flame deflector.

Other flush mounted nozzles, positioned around the pad,
will wash away any fluid spill as a protection against fire
hazards.

Water spray systems also are available along the
egress route that the astronauts and closeout crews would
follow in case an emergency evacuation was required.

Flame Trench and Deflector

The flame trench is 58 feet wide and approximately six
feet above mean sea level at the base. The height of the
trench and deflector is approximately 42 feet.

The flame deflector weighs about 1.3 million pounds and
is stored outside the flame trench on rails. When it is moved
beneath the launcher, it is raised hydraulically into position.
The deflector is covered with a four-and-one-half-inch thickness of refractory concrete consisting of a volcanic ash aggregate
and a calcuim aluminate binder. The heat and blast of the
engines are expected to wear about three-quarters of an inch
from this refractory surface during the Apollo 9 launch.

Pad Areas

Both Pad A and Pad B of Launch Complex 39 are roughly
octagonal in shape and cover about one fourth of a square
mile of terrain.

The center of the pad is a hardstand constructed of
heavily reinforced concrete. In addition to supporting the
weight of the mobile launcher and the Saturn V vehicle, it
also must support the 9.8-million-pound mobile service structure
and 6-million-pound transporter, all at the same time. The
top of the pad stands some 48 feet above sea level.

Saturn V propellants -- liquid oxygen, liquid hydrogen,
and RP-1 -- are stored near the pad perimeter.

Stainless steel, vacuum-jacketed pipes carry the liquid
oxygen (LOX) and liquid hydrogen from the storage tanks to
the pad, up the mobile launcher, and finally into the launch
vehicle propellant tanks.

LOX is supplied from a 900,000-gallon storage tank. A centrifugal pump with a discharge pressure of 320 pounds-per-square-inch pumps LOX to the vehicle at flow rates as high as 10,000-gallons-per-minute.

Liquid hydrogen, used in the second and third stages, is stored in an 850,000-gallon tank, and is sent through 1,500 feet of 10-inch, vacuum-jacketed invar pipe. A vaporizing heat exchanger pressurizes the storage tank to 60 psi for a 10,000-gallons-per-minute flow rate.

The RP-1 fuel, a high grade of kerosene is stored in three tanks--each with a capacity of 86,000 gallons. It is pumped at a rate of 2,000 gallons-per-minute at 175 psig.

The Complex 39 pneumatic system includes a converter-compressor facility, a pad high-pressure gas storage battery, a high-presssure storage battery in the VAB, low and high-pressure, cross-country supply lines, high-pressure hydrogen storage and conversion equipment, and pad distribution piping to pneumatic control panels. The various purging systems require 187,000 pounds of liquid nitrogen and 21,000 gallons of helium.

-more-

MISSION CONTROL CENTER

The Mission Control Center at the Manned Spacecraft
Center, Houston, is the focal point for all Apollo flight
control activities. The Center will receive tracking
and telemetry data from the Manned Space Flight Network.
These data will be processed through the Mission Control
Center Real-Time Computer Complex and used to drive
displays for the flight controllers and engineers in
the Mission Operations Control Room and staff support rooms.

The Manned Space Flight Network tracking and data
acquisition stations link the flight controllers at the
Center to the spacecraft.

For Apollo 9, all stations will be remote sites
without flight control teams. All uplink commands and
voice communications will originate from Houston, and
telemetry data will be sent back to Houston at high speed
(2,400 bits per second), on two separate data lines.
They can be either real time or playback information.

Signal flow for voice circuits between Houston and
the remote sites is via commercial carrier, usually
satellite, wherever possible using leased lines which
are part of the NASA Communications Network.

Commands are sent from Houston to NASA's Goddard
Space Flight Center, Greenbelt, Md., lines which link
computers at the two points. The Goddard computers
provide automatic switching facilities and speed buffering
for the command data. Data are transferred from Goddard
to remote sites on high speed (2,400 bits per second)
lines. Command loads also can be sent by teletype from
Houston to the remote sites at 100 words per minute.
Again, Goddard computers provide storage and switching
functions.

Telemetry data at the remote site are received by
the RF receivers, processed by the Pulse Code Modulation
ground stations, and transferred to the 642B remote-site
telemetry computer for storage. Depending on the format
selected by the telemetry controller at Houston, the
642B will output the desired format through a 2010 data
transmission unit which provides parallel to serial
conversion, and drives a 2,400 bit-per-second mode.

- more -

The data modem converts the digital serial data
to phase-shifted keyed tones which are fed to the high
speed data lines of the Communications Network.

Tracking data are output from the sites in a low
speed (100 words) teletype format and a 240-bit block high
speed (2,400 bits) format. Data rates are one sample-6 seconds for teletype and 10 samples (frames) per second for high
speed data.

All high-speed data, whether tracking or telemetry,
which originate at a remote site are sent to Goddard on high-
speed lines. Goddard reformats the data when necessary and
sends them to Houston in 600-bit blocks at a 40,800 bits-per-
second rate. Of the 600-bit block, 480 bits are reserved for
data, the other 120 bits for address, sync, intercomputer instru-
ctions, and polynominal error encoding.

All wideband 40,800 bits-per-second data originating at
Houston are converted to high speed (2,400 bits-per-second)
data at Goddard before being transferred to the designated remote site.

MANNED SPACE FLIGHT NETWORK

The Manned Space Flight Tracking Network for Apollo 9, consisting of 14 ground stations, four instrumented ships and six instrumented aircraft, is participating in its third manned flight. It is the global extension of the monitoring and control capability of the Mission Control Center in Houston. The network, developed by NASA through the Mercury and Gemini programs, now represents an investment of some $500 million and, during flight operations, has 4,000 persons on duty. In addition to NASA facilities, the network includes facilities of the Department of Defense and the Australian Department of Supply.

The network was developed by the Goddard Space Flight Center (GSFC) under the direction of NASA's Office of Tracking and Data Acquisition.

Basically, manned flight stations provide one or more of the following functions for flight control:

1. Telemetry
2. Tracking
3. Command and
4. Voice communications with the spacecraft

Apollo missions require the network to obtain information -- instantly recognize it, decode it, and arrange it for computer processing and display in the Mission Control Center.

Apollo generates much more information than either Projects Mercury or Gemini; therefore, very high speed data processing and display capability are needed. Apollo also requires network support at both Earth orbital and lunar distances. The Apollo Unified S-Band System (USB) provides this capability.

MANNED SPACE FLIGHT NETWORK APOLLO-9

Network Configuration for Apollo 9

Unified S-Band Sites

NASA 30-Ft. Antenna Sites	NASA 85-Ft. Antenna Sites

Antigua (ANG)
Ascension Island (ACN)
Bermuda (BDA)
Canary Island (CYI)
Carnarvon (CRO), Australia
Grand Bahama Island (GBM)
Guam (GWM)
Guaymas (GYM), Mexico
Hawaii (HAW)
Merritt Island (MIL), Fla.
Texas (TEX), Corpus Christi

Honeysuckle Creek (HSK),
   Australia (Prime)
Goldstone (GDS), Calif.
   (Prime)
Madrid (MAD), Spain (Prime)
*Canberra (DSS-42-Apollo Wing)
   (Backup)
*Goldstone (DSS-11-Apollo Wing)
   (Backup) MARS 210'-(Backup)
*Madrid (DSS-61-Apollo Wing)
   (Backup)

Tananarive (TAN), Malagasy Republic (STADAN station in support
role only.)

* Wings have been added to JPL Deep Space Network site operations buildings. These wings contain additional Unified S-Band equipment as backup to the Prime sites to allow for the use, if necessary, of the Deep Space 85-ft. antennas.

Network Testing

The MSFN began its Network Readiness Testing for Apollo 9 on Jan. 20 and continued through Feb. 5 when the network went on mission status. Through established computer techniques Goddard's Real-Time Computer Center (RTCC) conducted system-by-system, station-by-station tests of all tracking/data acquisition and communications systems until all support criteria were met and the MSFN pronounced ready to participate in the mission.

Spacecraft Communications

All Manned Space Flight Network stations are prepared to communicate with the Apollo 9 spacecraft in two different modes, S-Band and VHF.

When a station acquires the spacecraft, those sites having Unified S-Band and VHF air-to-ground capability will select the best quality and pass it to the Mission Control Center. All stations will monitor air-to-ground conversations for a possible crew request to switch the spacecraft communications.

NASA Communications Network - Goddard

This network consists of several systems of diversely routed communications channels leased on communications satellies, common carrier systems and high frequency radio facilities where necessary to provide the access links.

The system consists of both narrow and wide-band channels, and some TV channels. Included are a variety of telegraph, voice and data systems (digital and analog) with a wide range of digital data rates. Alternate routes or redundancy are provided for added reliability.

A primary switching center and intermediate switching and control points are established to provide centralized facility and technical control under direct NASA control. The primary switching center is at Goddard, and intermediate switching centers are located at Canberra, Australia; Madrid, Spain; London, England; Honolulu, Hawaii; Guam and Cape Kennedy, Fla.

Cape Kennedy is connected directly to the Mission Control Center by the communication network's Apollo Launch Data System (ALDS), a combination of data gathering and transmission systems designed to handle launch data exclusively.

After launch all network and tracking data are directed to the Mission Control Center through Goddard. A high-speed data line connects Cape Kennedy to Goddard, where the transmission rate is increased from there to Mission Control Center. Upon orbital insertion, tracking responsibility is transferred between the various stations as the spacecraft circles the Earth.

Two Intelsat communications satellites will be used for Apollo 9, one positioned over the Atlantic Ocean at about 60 degrees W. longitude in a near equatorial synchronous orbit varying about six degrees N. and S. in latitude. The Atlantic satellite will service the Ascension Island USB station, the Atlantic Ocean ship and the Canary Island site.

Only two of these three stations will be transmitting information back to Goddard at any one time, but all three stations can receive at all times.

The second Apollo Intelsat communications satellite is located about 170 degrees E. longitude over the mid-Pacific near the Equator at the international dateline. It will service the Carnarvon, Australian USB site and the Pacific Ocean ships. All these stations will be able to transmit simultaneously through the satellites to the Mission Control Center via Jamesburg, Calif., and the Goddard Space Flight Center.

NASCOM—APOLLO 9

Sites with "Dual" Capability

Certain stations of the Manned Space Flight Network can provide tracking, voice and data acquisition for two Apollo spacecraft simultaneously, provided they are within the beamwidth of the single Unified S-Band antenna. Two sets of frequencies separated by approximately five megahertz are used for this purpose. In addition to this primary mode of communications, the Unified S-Band system has the capability of receiving data on two other frequencies; primarily used for downlink data from the CSM.

For Apollo 9, this capability will be utilized when the LM and CSM have separated. Effective "dual" acquisition displacement distance between CSM and LM is one-hundred miles at a relative angle of forty-five degrees.

The "Dual" sites:

85 Ft. Antenna Systems

Honeysuckle Creek, Australia	Prime Site Wing Site (backup)
Madrid, Spain	Prime Site Wing Site (backup)
Goldstone, Calif.	Prime Site Wing Site (backup)

30 Ft. Antenna Systems

Carnarvon, Australia	Kauai, Hawaii
Ascension Island	Bermuda (Uplink only)
Merritt Island, Fla.	Antigua (Uplink only)
Guam Island, Pacific	USNS REDSTONE USNS MERCURY USNS VANGUARD

Spacecraft Television

Television transmissions will be received, recorded and converted to commercial (home) format for release to the public by the Merritt Island, Fla., and Goldstone, Calif., Manned Space Flight Network stations.

-more-

Land Station Tracking

After the S-IVB and CSM/LM separate, dual-capability stations will be required to track both vehicles simultaneously to provide HF A/G voice remoting from the CSM/LM and VHF TLM data from the S-IVB/IU. After LM power up, dual capability stations will be required to track the CSM via acquisition bus to provide VHF voice, and to track the LM on VHF to provide VHF TLM or voice. Stations having only one VHF system will track the CSM or LM in accordance with Houston requirements.

SHIPS AND AIRCRAFT NETWORK SUPPORT - APOLLO 9

The Apollo Instrumentation Ships (AIS)

The Apollo 9 mission will be supported by four Apollo Instrumentation Ships operating as integral stations of the Manned Space Flight Network (MSFN) to provide coverage in areas beyond the range of land stations. The ships USNS Vanguard, Redstone, Mercury, and Huntsville will perform tracking, telemetry, communications, and computer functions for the launch and Earth orbit insertion phase, LM maneuver phases, and re-entry at end of mission.

The Vanguard will be stationed due east of Bermuda (32 degrees N - 45 degrees W) during launch and will remain at this position to cover specific orbital phases. The Vanguard also functions as part of the Atlantic recovery fleet in the event of a launch phase contingency. The Redstone, Mercury, and Huntsville will all be deployed in the Pacific Ocean area. The Redstone will be east of Hawaii (22 degrees N - 131 degrees W), the Mercury southeast of American Samoa (22 degrees S - 160 degrees W), and the Huntsville northwest of American Samoa (7 degrees S - 170 degrees E). The stations will provide communications, tracking, and telemetry coverage in the areas where spacecraft and LM functions occur without adequate coverage from land stations. Since the ships are a mobile instrumentation station, they can be repositioned prior to or during the mission based upon flight plan needs subject to their speed limitations and weather conditions.

The Apollo ships were developed jointly by NASA and the Department of Defense. The DOD operates the ships in support of Apollo and other NASA/DOD missions on a noninterference basis with Apollo support requirements. The overall management of the Apollo ship operation is the responsibility of the Air Force Western Test Range (AFWTR). The Military Sea Transport Service (MSTS) provides the maritime crews; the Federal Electric Co. (FEC) (under contract to AFWTR) provides the technical instrumentation crews. Goddard Space Flight Center (GSFC) is responsible for the configuration control and network interface of the ships in support of Apollo missions. The technical crews operate in accordance with joint NASA/DOD technical standards and specifications which are compatible with the manned space flight network procedures.

The Apollo Range Instrumentation Aircraft (ARIA)

The ARIA will support the Apollo 9 mission by filling gaps in critical coverage beyond the range of land and ship stations or at points where land or ship coverage is either impossible or impractical. During this mission six ARIA will be used in the Pacific and Atlantic areas to receive and record telemetry from the S-IVB stage, and later on in the mission from the CSM/LM. In addition, the ARIA will provide a two-way voice relay between the Mission Control Center and the Apollo 9 crew during critical maneuver periods.

On launch day, four ARIA will be used in the Pacific area during revolutions 2, 3, and 4 to cover LM/CSM/S-IVB functions. These aircraft will use staging bases in Australia, Guam, and Hawaii. On day two, revolution 27, one ARIA will deploy to a point about 400 miles ENE of Darwin, Australia, and a second ARIA will deploy to a point some 900 miles SSE of Tahiti to provide critical telemetry and voice relay coverage during activiation of LM electrical propulsion and environmental control systems. Staging bases for this coverage will be Darwin and Pago Pago. On day four, two ARIA will be deployed in the Atlantic area northeast and northwest of Ascension Island to provide active voice relay at predetermined points on revolutions 61, 62, and 63 during CSM/LM maneuvers. At the end of mission, at least one ARIA will be positioned south of Bermuda to provide telemetry and voice relay from the CSM from the 400K foot level to splashdown. ARIA operating in the Atlantic area will use bases at Ascension Island, Puerto Rico, and Patrick AFB, Florida.

The total ARIA fleet consists of eight EC-135N (Boeing 707) jet aircraft equipped specifically to meet mission needs. Seven foot diameter parabolic (dish) antennas have been installed in the nose section of each aircraft giving them a large, bulbous look. The ARIA airframes were selected from the USAF transport fleet, modified through joint DOD/NASA contract, and are operated and maintained by the Air Force Eastern Test Range (AFETR). During their support of Apollo missions, they become an integral part of the MSFN and, as such, operate in accordance with the joint NASA/DOD procedures and policies which govern the MSFN.

APOLLO 9 CREW

Crew Training

The crewmen of Apollo 9 will have spent more than seven hours of formal training for each hour of the mission's 10-day duration. Almost 1,800 hours of training were in the Apollo 9 crew training syllabus over and above the normal preparations for the mission -- technical briefings and reviews, pilot meetings and study.

The Apollo 9 crewmen also took part in spacecraft manufacturing checkouts at the North American Rockwell plant in Downey, Calif., at Grumman Aircraft Engineering Corp., Bethpage, N. Y., and in prelaunch testing at NASA Kennedy Space Center. Taking part in factory and launch area testing has provided the crew with thorough operational knowledge of the complex vehicle.

Highlights of specialized Apollo 9 crew training topics are:

* Detailed series of briefings on spacecraft systems, operation and modifications.

* Saturn launch vehicle briefings on countdown, range safety, flight dynamics, failure modes and abort conditions. The launch vehicle briefings were updated periodically.

* Apollo Guidance and Navigation system briefings at the Massachusetts Institute of Technology Instrumentation Laboratory.

* Briefings and continuous training on mission photographic objectives and use of camera equipment.

* Extensive pilot participation in reviews of all flight procedures for normal as well as emergency situations.

* Stowage reviews and practice in training sessions in the spacecraft, mockups, and Command Module simulators allowed the crewmen to evaluate spacecraft stowage of crew-associated equipment.

* Zero-g aircraft flights using command module and lunar module mockups for EVA and pressure suit doffing/donning practice and training.

* Underwater zero-g training in the MSC Water Immersion Facility using spacecraft mockups to further familiarize crew with all aspects of CSM-LM docking tunnel intravehicular transfer and EVA in pressurized suits.

* More than 300 hours of training per man in command module and lunar module mission simulators at MSC and KSC, including closed-loop simulations with flight controllers in the Mission Control Center. Other Apollo simulators at various locations were used extensively for specialized crew training.'

* Water egress training conducted in indoor tanks as well as in the Gulf of Mexico, included uprighting from the Stable II position (apex down) to the Stable I position (apex up), egress onto rafts and helicopter pickup.

* Launch pad egress training from mockups and from the actual spacecraft on the launch pad for possible emergencies such as fire, contaminants and power failures.

* The training covered use of Apollo spacecraft fire suppression equipment in the cockpit.

* Planetarium reviews at Morehead Planetarium, Chapel Hill, N. C., and at Griffith Planetarium, Los Angeles, of the celestial sphere with special emphasis on the 37 navigational stars used by the Command Module Computer.

Crew Life Support Equipment

Apollo 9 crewmen will wear two versions of the Apollo spacesuit: an intravehicular pressure garment assembly worn by the command module pilot and the extravehicular pressure garment assembly worn by the commander and the lunar module pilot. Both versions are basically identical except that the extravehicular version has an integral thermal meteoroid garment over the basic suit.

From the skin out, the basic pressure garment consists of a nomex comfort layer, a neoprene-coated nylon pressure bladder and a nylon restraint layer. The outer layers of the intravehicular suit are, from the inside out, nomex and two layers of Teflon-coated Beta cloth. The extravehicular integral thermal meteoroid cover consists of a liner of two layers of neoprene-coated nylon, seven layers of Beta/Kapton spacer laminate, and an outer layer of Teflon-coated Beta fabric.

The extravehicular suit, together with a liquid cooling garment, portable life support system (PLSS), oxygen purge system, extravehicular visor assembly and other components make up the extravehicular mobility unit (EMU). The EMU provides an extravehicular crewman with life support for a four-hour mission outside the lunar module without replenishing expendables. EMU total weight is 183 pounds. The intravehicular suit weighs 35.6 pounds.

Liquid cooling garment--A knitted nylon-spandex garment with a network of plastic tubing through which cooling water from the PLSS is circulated. It is worn next to the skin and replaces the constant wear garment during EVA only.

Portable life support system--A backpack supplying oxygen at 3.9 psi and cooling water to the liquid cooling garment. Return oxygen is cleansed of solid and gas contaminants by a lithium hydroxide canister. The PLSS includes communications and telemetry equipment, displays and controls and a main power supply. The PLSS is covered by a thermal insulation jacket.

Oxygen purge system--Mounted atop the PLSS, the oxygen purge system provides a contingency 30-minute supply of gaseous oxygen in two two-pound bottles pressurized to 5,880 psia. The system may also be worn separately on the front of the pressure garment assembly torso. It serves as a mount for the VHF antenna for the PLSS.

Extravehicular visor assembly--A polycarbonate shell and two visors with thermal control and optical coatings on them. The EVA visor is attached over the pressure helmet to provide impact, micrometeoroid, thermal and light protection to the EVA crewman.

Extravehicular gloves--Built of an outer shell of Chromel-R fabric and thermal insulation to provide protection when handling extremely hot and cold objects. The finger tips are made of silicone rubber to provide the crewman more sensitivity.

A one-piece constant wear garment, similar to "long johns", is worn as an undergarment for the spacesuit in intravehicular operations and for the inflight coveralls. The garment in porous-knit cotton with a waist-to-neck zipper for donning. Biomedical harness attach points are provided.

During periods out of the spacesuits, crewmen will wear two-piece Teflon fabric inflight coveralls for warmth and for pocket stowage of personal items.

Communications carriers ("Snoopy hats") with redundant microphones and earphones are worn with the pressure helmet; a lightweight headset is worn with the inflight coveralls.

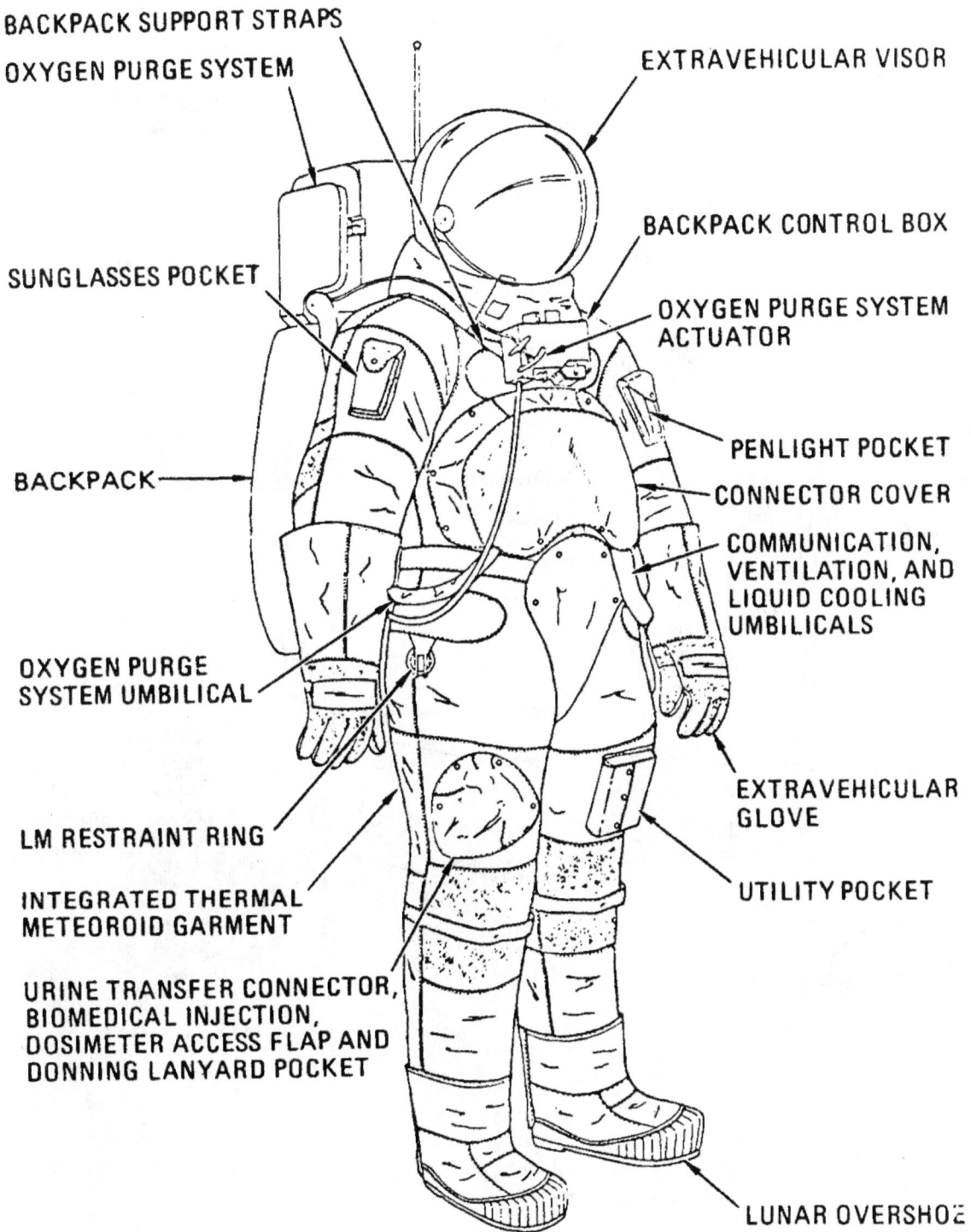

BACKPACK SUPPORT STRAPS

OXYGEN PURGE SYSTEM

EXTRAVEHICULAR VISOR

SUNGLASSES POCKET

BACKPACK CONTROL BOX

OXYGEN PURGE SYSTEM ACTUATOR

BACKPACK

PENLIGHT POCKET

CONNECTOR COVER

COMMUNICATION, VENTILATION, AND LIQUID COOLING UMBILICALS

OXYGEN PURGE SYSTEM UMBILICAL

EXTRAVEHICULAR GLOVE

LM RESTRAINT RING

UTILITY POCKET

INTEGRATED THERMAL METEOROID GARMENT

URINE TRANSFER CONNECTOR, BIOMEDICAL INJECTION, DOSIMETER ACCESS FLAP AND DONNING LANYARD POCKET

LUNAR OVERSHOE

P-269

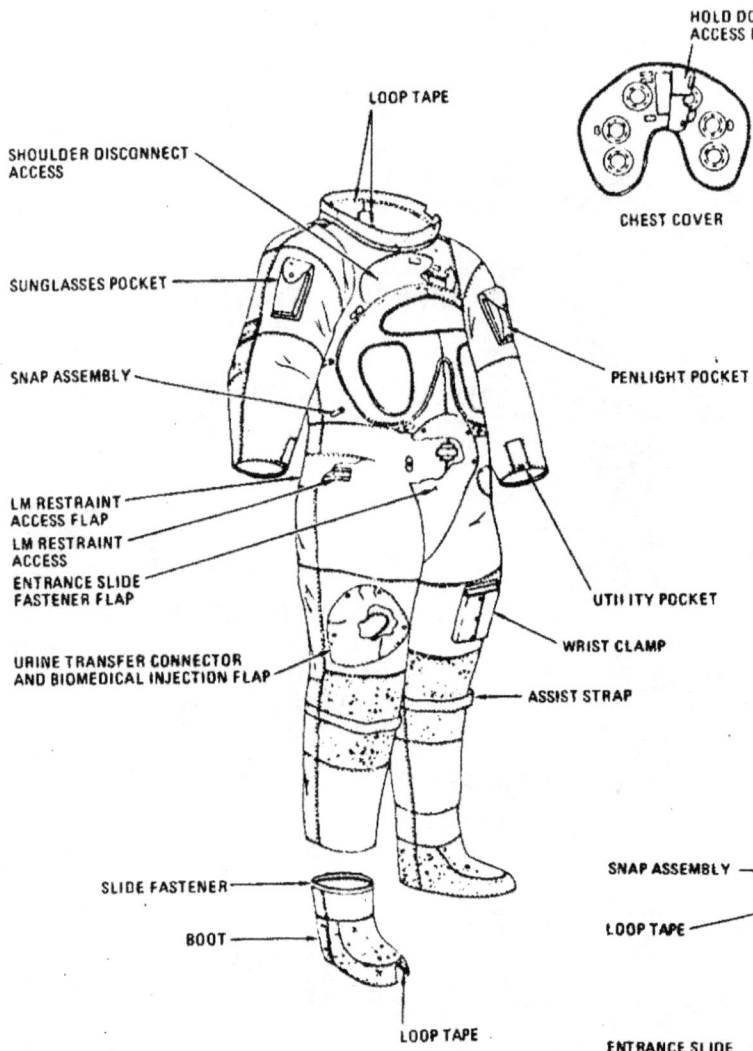

HOLD DOWN STRAP
ACCESS FLAP

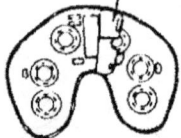

CHEST COVER

CONNECTOR COVER

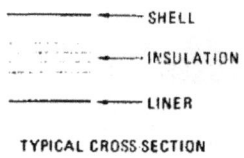

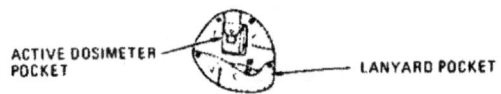

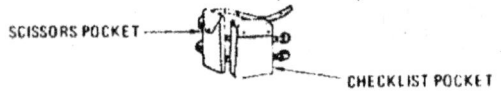

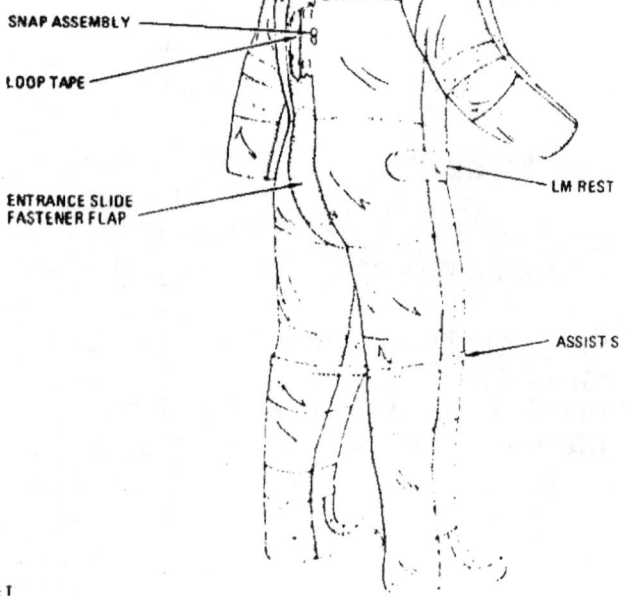

LOOP TAPE

SHOULDER DISCONNECT
ACCESS

SUNGLASSES POCKET

SNAP ASSEMBLY

PENLIGHT POCKET

SHELL
INSULATION
LINER

TYPICAL CROSS SECTION

LM RESTRAINT
ACCESS FLAP

LM RESTRAINT
ACCESS

ENTRANCE SLIDE
FASTENER FLAP

URINE TRANSFER CONNECTOR
AND BIOMEDICAL INJECTION FLAP

UTILITY POCKET

WRIST CLAMP

ASSIST STRAP

BELT ASSEMBLY

DATA LIST POCKET

SLIDE FASTENER

BOOT

LOOP TAPE

SNAP ASSEMBLY

LOOP TAPE

ENTRANCE SLIDE
FASTENER FLAP

LM REST

ASSIST S

ACTIVE DOSIMETER
POCKET

LANYARD POCKET

SCISSORS POCKET

CHECKLIST POCKET

P-270

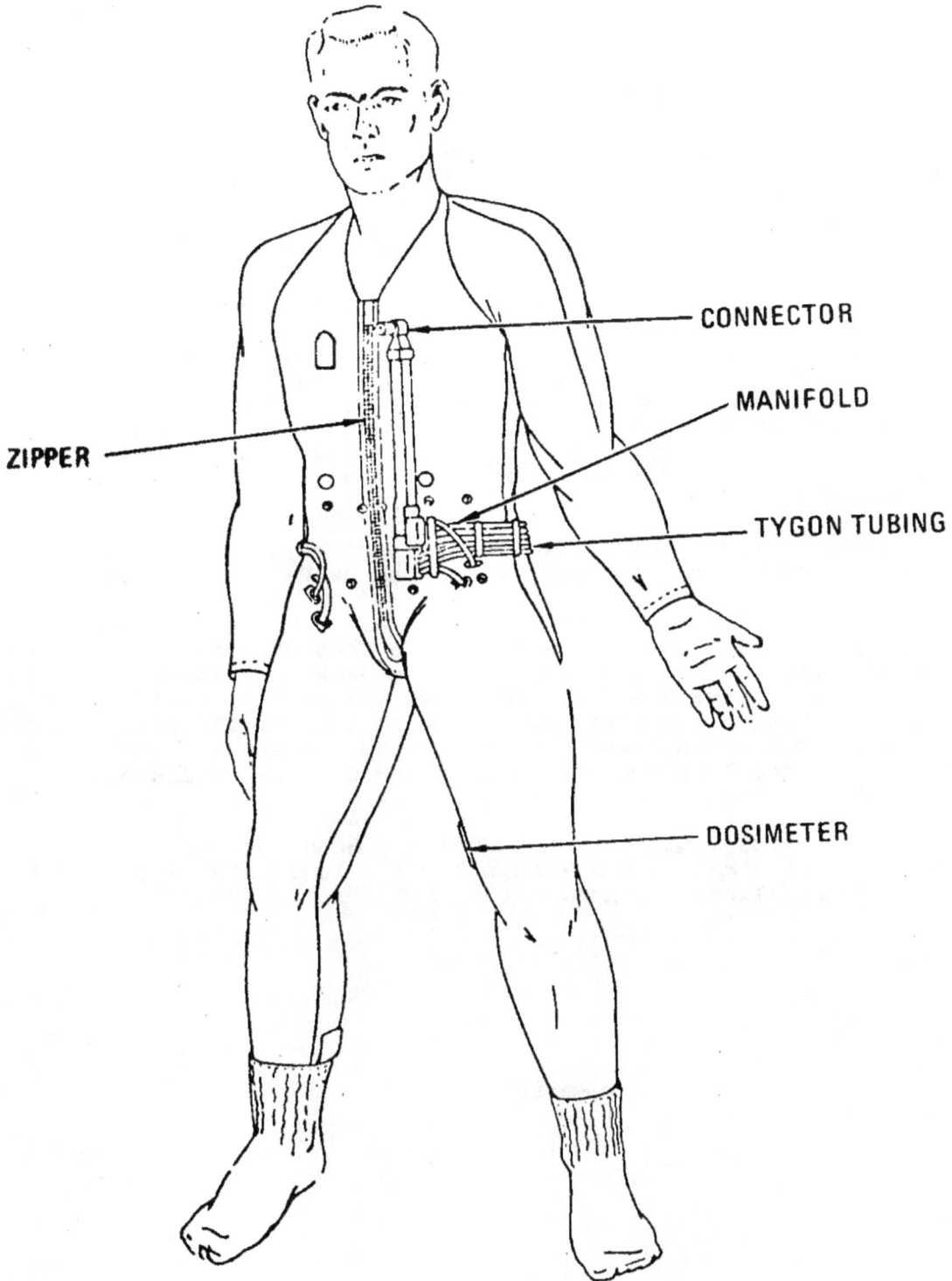

CONNECTOR

MANIFOLD

ZIPPER

TYGON TUBING

DOSIMETER

P-274

Apollo 9 Crew Meals

The Apollo 9 crew has a wide range of food items from which to select their daily mission space menu. More than 60 items comprise the food selection list of freeze-dried bite-size and rehydratable foods. The average daily value of three meals will be 2,500 calories per man.

Water for drinking and rehydrating food is obtained from three sources in the command module -- a dispenser for drinking water and two water spigots at the food preparation station, one supplying water at about 155 degrees F, the other at about 55 degrees F. The potable water dispenser emits half-ounce spurts with each squeeze and the food preparation spigots dispense water in 1-ounce increments. Command module potable water is supplied from service module fuel cell by-product water.

A similar hand water dispenser aboard the lunar module is used for cold-water rehydration of food packets stowed in the LM.

After water has been injected into a food bag, it is kneaded for about three minutes. The bag neck is then cut off and the food squeezed into the crewman's mouth. After a meal, germicide pills attached to the outside of the food bags are placed in the bags to prevent fermentation and gas formation and the bags are rolled and stowed in waste disposal compartments.

The day-by-day, meal-by-meal Apollo 9 menu for each crewman for both the command module and the lunar module is listed on the following pages.

APOLLO 9 - McDivitt

MEAL	DAY 1*, 5, 9	DAY 2, 6, 10	DAY 3, 7, 11	DAY 4, 8
A	Peaches Bacon Squares (8) Cinn Tstd Bread Cubes (8) Grapefruit Drink Orange Drink	Canadian Bacon & Applesauce Sugar Coated Corn Flakes Brownies (8) Grapefruit Drink Grape Drink	Fruit Cocktail Bacon Squares (8) Cinn Tstd Bread Cubes (8) Cocoa Orange Drink	Sausage Patties Peaches Bacon Squares (8) Cocoa Grape Drink
B	Salmon Salad Chicken & Gravy Toasted Bread Cubes (6) Sugar Cookie Cubes (6) Cocoa	Tuna Salad Chicken & Vegetables Cinn Tstd Bread Cubes (8) Pineapple Fruitcake (4) Pineapple-Grapefruit Drink	Cream of Chicken Soup Beef Pot Roast Toasted Bread Cubes (8) Butterscotch Pudding Grapefruit Drink	Pea Soup Chicken & Gravy Cheese Sandwiches (6) Bacon Squares (6) Grapefruit Drink
C	Beef & Gravy Beef Sandwiches (4) Cheese-Cracker Cubes (8) Chocolate Pudding Orange-Grapefruit Drink	Spaghetti & Meat Sauce Beef Bites (6) Bacon Squares (6) Banana Pudding Grapefruit Drink	Beef Hash Chicken Salad Turkey Bites (6) Graham Cracker Cubes (6) Orange Drink	Shrimp Cocktail Beef & Vegetables Cinn Tstd Bread Cubes (8) Date Fruitcake (4) Orange-Grapefruit Drink

*Day 1 consists of Meals B and C only
Each crewmember will be provided with a total of 32 meals

APOLLO 9 - Scott

MEAL	DAY 1*, 5, 9	DAY 2, 6, 10	DAY 3, 7, 11	DAY 4, 8
A	Peaches Bacon Squares (8) Cinn Tstd Bread Cubes (8) Grapefruit Drink Orange Drink	Canadian Bacon & Applesauce Sugar Coated Corn Flakes Brownies (8) Grapefruit Drink Grape Drink	Fruit Cocktail Bacon Squares (8) Cinn Tstd Bread Cubes (8) Cocoa Orange Drink	Sausage Patties Peaches Bacon Squares (8) Cocoa Grape Drink
B	Salmon Salad Chicken & Gravy Toasted Bread Cubes (6) Sugar Cookie Cubes (6) Cocoa	Tuna Salad Chicken & Vegetables Cinn Tstd Bread Cubes (8) Pineapple Fruitcake (4) Pineapple-Grapefruit Drink	Cream of Chicken Soup Beef Pot Roast Toasted Bread Cubes (8) Butterscotch Pudding Grapefruit Drink	Pea Soup Chicken & Gravy Cheese Sandwiches (6) Bacon Squares (6) Grapefruit Drink
C	Beef & Gravy Beef Sandwiches (4) Cheese-Cracker Cubes (8) Chocolate Pudding Orange-Grapefruit Drink	Spaghetti & Meat Sauce Beef Bites (6) Bacon Squares (6) Banana Pudding Grapefruit Drink	Beef Hash Chicken Salad Turkey Bites (6) Graham Cracker Cubes (6) Orange Drink	Shrimp Cocktail Beef & Vegetables Cinn Tstd Bread Cubes (8) Date Fruitcake (4) Orange-Grapefruit Drink

*Day 1 consists of Meals B and C only
Each crewmember will be provided with a total of 32 meals

-more-

APOLLO 9 - Schweickart

MEAL	DAY 1*, 5, 9	DAY 2, 6, 10	DAY 3, 7, 11	DAY 4, 8
A	Peaches Bacon Squares (8) Cinn Tstd Bread Cubes (8) Grapefruit Drink Orange Drink	Canadian Bacon & Applesauce Sugar Coated Corn Flakes Brownies (8) Grapefruit Drink Grape Drink	Fruit Cocktail Bacon Squares (8) Cinn Tstd Bread Cubes (8) Cocoa Orange Drink	Sausage Patties Peaches Bacon Squares (8) Cocoa Grape Drink
B	Salmon Salad Chicken & Gravy Toasted Bread Cubes (6) Sugar Cookie Cubes (6) Cocoa	Tuna Salad Chicken & Vegetables Cinn Tstd Bread Cubes (8) Pineapple Fruitcake (4) Pineapple-Grapefruit Drink	Cream of Chicken Soup Beef Pot Roast Toasted Bread Cubes (8) Butterscotch Pudding Grapefruit Drink	Pea Soup Chicken & Gravy Cheese Sandwiches (6) Bacon Squares (6) Grapefruit Drink
C	Beef & Gravy Beef Sandwiches (4) Cheese-Cracker Cubes (8) Chocolate Pudding Orange-Grapefruit Drink	Spaghetti & Meat Sauce Beef Bites (6) Bacon Squares (6) Banana Pudding Grapefruit Drink	Beef Hash Chicken Salad Turkey Bites (6) Graham Cracker Cubes (6) Orange Drink	Spaghetti & Meat Sauce Beef & Vegetables Cinn Tstd Bread Cubes (8) Date Fruitcake (4) Orange-Grapefruit Drink

*Day 1 consists of Meals B and C only
Each crewmember will be provided with a total of 32 meals

APOLLO 9 LM MENU

Day 1

Meal A

Chicken and Gravy
Butterscotch Pudding
Sugar Cookie Cubes (6)
Orange-Pineapple Drink
Grape Drink

Meal B

Chicken Salad
Beef Sandwiches (6)
Date Fruitcake (4)
Chocolate Pudding
Orange Drink

Meal C

Beef Hash
Bacon Squares (8)
Strawberry Cereal Cubes (6)
Pineapple-Grapefruit Drink
Grape Drink

2 man-days are required
Red and Blue Velcro
2 meals per overwrap
1-10-69

-more-

Personal Hygiene

Crew personal hygiene equipment aboard Apollo 9 includes body cleanliness items, the waste management system and two medical kits.

Packaged with the food are a toothbrush and a 2-ounce tube of toothpaste for each crewman. Each man-meal package contains a 3.5 by 4-inch wet-wipe cleansing towel. Additionally, three packages of 12 by 12-inch dry towels are stowed beneath the command module pilot's couch. Each pakcage contains seven towels. Also stowed under the command module pilot's couch are seven tissue dispensers containing 53 3-ply tissues each.

Solid body wastes are collected in Gemini-type plastic defecation bags which contain a germicide to prevent bacteria and gas formation. The bags are sealed after use and stowed in empty food containers for post-flight analysis.

Urine collection devices are provided for use either wearing the pressure suit while in the inflight coveralls. The urine is dumped overboard through the spacecraft urine dump valve in the CM and stored in the LM.

The two medical accessory kits, 6 by 4.5 by 4 inches, are stowed on the spacecraft back wall at the feet of the command module pilot.

The medical kits contain three motion sickness injectors, three pain suppression injectors, one 2-ounce bottle first aid ointment, two 1-ounce bottle eye drops, three nasal sprays, two compress bandages, 12 adhesive bandages, one oral thermometer and two spare crew biomedical harnesses. Pills in the medical kits are 60 antibiotic, 12 nausea, 12 stimulant, 18 pain killer, 60 decongestant, 24 diarrhea, 72 aspirin and 21 sleeping. Additionally, a small medical kit containing two stimulant and four pain killer pills is stowed in the lunar module food compartment.

Survival Gear

The survival kit is stowed in two rucksacks in the right-hand forward equipment bay above the lunar module pilot.

Contents of rucksack No. 1 are: two combination survival lights, one desalter kit, three pair sunglasses, one radio beacon, one spare radio beacon battery and spacecraft connector cable, one knife in sheath, three water containers and two containers of Sun lotion.

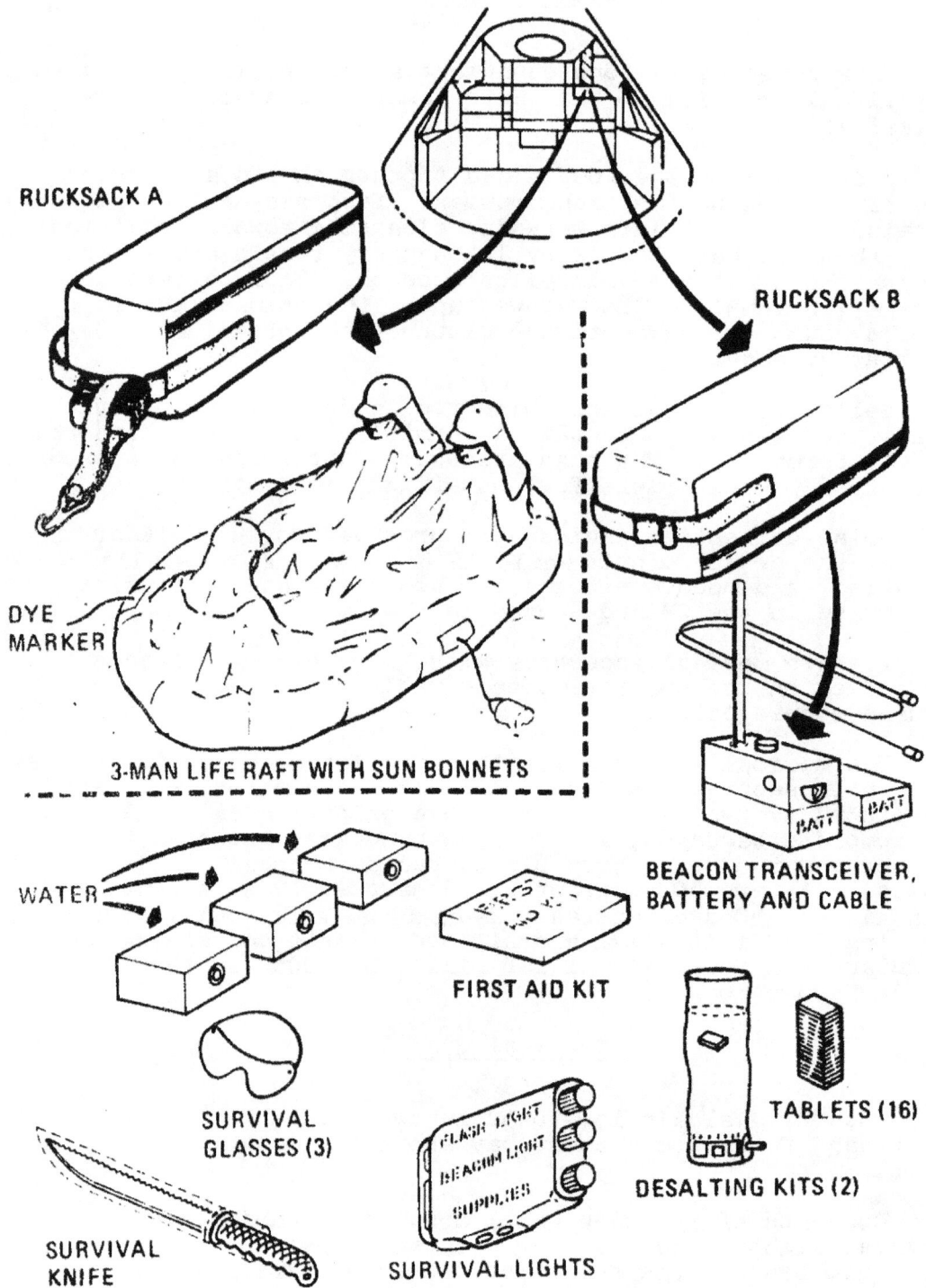

RUCKSACK A

RUCKSACK B

DYE
MARKER

3-MAN LIFE RAFT WITH SUN BONNETS

BEACON TRANSCEIVER,
BATTERY AND CABLE

WATER

FIRST AID KIT

SURVIVAL
GLASSES (3)

TABLETS (16)

DESALTING KITS (2)

SURVIVAL
KNIFE

FLASH LIGHT
BEACON LIGHT
SUPPLIES

SURVIVAL LIGHTS

Rucksack No. 2: one three-man life raft with CO_2 inflater, one sea anchor, two sea dye markers, three sunbonnets, one mooring lanyard, three manlines and two attach brackets.

The survival kit is designed to provide a 48-hour postlanding (water or land) survival capability for three crewmen between 40 degrees North and South latitudes.

Biomedical Inflight Monitoring

The Apollo 9 crew biomedical telemetry data received by the Manned Space Flight Network will be relayed for instantaneous display at Mission Control Center where heart rate and breathing rate data will be displayed on the flight surgeon's console. Heart rate and respiration rate average, range and deviation are computed and displayed on digital TV screens.

In addition, the instantaneous heart rate, real time and delayed EKG and respiration are recorded on strip charts for each man.

Biomedical telemetry will be simultaneous from all crewmen while in the CSM, but selectable by a manual onboard switch in the LM. During EVA the flight surgeons will be able to monitor the EVA LM pilot as well as the commander in the LM and the command module pilot in the CSM.

Biomedical data observed by the MOCR flight surgeon and his team in the Life Support Systems Staff Support Room will be correlated with spacecraft and spacesuit environmental data displays.

Blood pressures are no longer telemetered as they were in the Mercury and Gemini programs. Oral temperature, however, can be measured onboard for diagnostic purposes and voiced down by the crew in case of inflight illness.

Crew Launch-Day Timeline

Following is a timetable of Apollo 9 crew activities on launch day. (All times are shown in hours and minutes before liftoff.)

T-9:00 - Backup crew alerted

T-8:30 - Backup crew to LC-39A for spacecraft prelaunch checkouts

T-5:00 - Flight crew alerted

T-4:45 - Medical examinations

T-4:15 - Breakfast

T-3:45 - Don pressure suits

T-3:30 - Leave Manned Spacecraft Operations Building for LC-39A via crew transfer van

T-3:14 - Arrive at LC-39A

T-3:10 - Enter elevator to spacecraft level

T-2:40 - Begin spacecraft ingress

Rest-Work Cycles

All three Apollo 9 crewmen will sleep simultaneously during rest periods. The commander and the command module pilot will sleep in the couches and the lunar module pilot will sleep in the lightweight sleeping bag beneath the couches. Since each day's mission activity is of variable length, rest periods will not come at regular intervals.

During rest periods, both the commander and the command module pilot will wear their communications headsets and remain on alert duty, but with receiver volume turned down.

When possible, all three crewmen will eat together in 1-hour eat periods during which other activities will be held a minimum.

-more-

Crew Biographies

NAME: James A. McDivitt (Colonel, USAF)
Spacecraft Commander

BIRTHPLACE AND DATE: Born June 10, 1929, in Chicago, Ill.
His parents, Mr. and Mrs. James McDivitt, reside in
Jackson, Mich.

PHYSICAL DESCRIPTION: Brown hair; blue eyes; height: 5 feet
11 inches; weight: 155 pounds.

EDUCATION: Graduated from Kalamazoo Central High School,
Kalamazoo, Mich.; received a Bachelor of Science degree
in Aeronautical Engineering from the University of
Michigan (graduated first in class) in 1959 and an
Honorary Doctorate in Astronautical Science from the
University of Michigan in 1965.

MARITAL STATUS: Married to the former Patricia A. Haas of
Cleveland, O. Her parents, Mr. and Mrs. William Haas,
reside in Cleveland.

CHILDREN: Michael A., Apr. 14, 1957; Ann L., July 21, 1958;
Patrick W., Aug. 30, 1960; Kathleen M., June 16, 1966.

OTHER ACTIVITIES: His hobbies include handball, hunting, golf,
swimming, water skiing and boating.

ORGANIZATIONS: Member of the Society of Experimental Test Pilots,
the American Institute of Aeronautics and Astronautics,
Tau Beta Pi and Phi Kappa Phi.

SPECIAL HONORS: Awarded the NASA Exceptional Service Medal and
the Air Force Astronaut Wings; four Distinguished Flying
Crosses; five Air Medals; the Chong Moo Medal from South
Korea; the USAF Air Force Systems Command Aerospace Primus
Award; the Arnold Air Society JFK Trophy; the Sword of
Loyola; and the Michigan Wolverine Frontiersman Award.

EXPERIENCE: McDivitt joined the Air Force in 1951 and holds the
rank of Colonel. He flew 145 combat missions during the
Korean War in F-80s and F-86s.

He is a graduate of the USAF Experimental Test Pilot
School and the USAF Aerospace Research Pilot course and
served as an experimental test pilot at Edwards Air Force
Base, Calif.

He has logged 3,922 hours of flying time -- 3,156 hours in
jet aircraft.

Colonel McDivitt was selected as an astronaut by NASA
in September 1962.

He was command pilot for Gemini 4, a 66-orbit 4-day
mission that began on June 3 and ended on June 7, 1965.
Highlights of the mission included a controlled extravehicular activity period performed by pilot Ed White,
cabin depressurization and opening of spacecraft cabin
doors, and the completion of 12 scientific and medical
experiments.

-more-

NAME: David R. Scott (Colonel, USAF)
       Command Module Pilot

BIRTHPLACE AND DATE: Born June 6, 1932, in San Antonio, Tex.
       His parents, Brigadier Gen. (USAF Ret.) and Mrs. Tom W.
       Scott, reside in LaJolla, Calif.

PHYSICAL DESCRIPTION: Blond hair; blue eyes; height: 6 feet;
       weight: 175 pounds.

EDUCATION: Graduated from Western High School, Washington, D.C.;
       received a Bachelor of Science degree from the U. S.
       Military Academy and the degree of Master of Science
       in Aeronautics and Astronautics from the Massachusetts
       Institute of Technology.

MARITAL STATUS: Married to the former Ann Lurton Ott of San
       Antonio, Tex. Her parents are Brigadier Gen. (USAF Ret.)
       and Mrs. Isaac W. Ott of San Antonio.

CHILDREN: Tracy L., Mar. 25, 1961; Douglas W., Oct. 8, 1963.

OTHER ACTIVITIES: His hobbies are swimming, handball, skiing
       and photography.

ORGANIZATIONS: Associate Fellow of the American Institute of
       Aeronautics and Astronautics; member of the Society of
       Experimental Test Pilots; Tau Beta Pi; Sigma Xi; and
       Sigma Gamma Tau.

SPECIAL HONORS: Awarded the NASA Exceptional Service Medal,
       the Air Force Astronaut Wings, and the Distinguished
       Flying Cross; and recipient of the AIAA Astronautics
       Award.

EXPERIENCE: Scott graduated fifth in a class of 633 at West
       Point and subsequently chose an Air Force career. He
       completed pilot training at Webb Air Force Base, Tex.,
       in 1955 and then reported for gunnery training at
       Laughlin Air Force Base, Tex., and Luke Air Force Base,
       Ariz.

       He was assigned to the 32nd Tactical Fighter Squadron at
       Soesterberg Air Force Base (RNAF), Netherlands, from
       Apr. 1956 to July 1960. Upon completing this tour of
       duty, he returned to the United States for study at the
       Massachusetts Institute of Technology where he completed
       work on his Master's degree. His thesis at MIT con-
       cerned interplanetary navigation.

After completing his studies at MIT in June 1962, he attended the Air Force Experimental Test Pilot School and then the Aerospace Research Pilot School.

He has logged more than 3,800 hours flying time -- 3,600 hours in jet aircraft.

Col. Scott was one of the third group of astronauts named by NASA in Oct. 1963.

On Mar. 16, 1966, he and command pilot Neil Armstrong were launched into space on the Gemini 8 mission -- a flight originally scheduled to last three days but terminated early due to a malfunctioning OAMS thruster. The crew performed the first successful docking of two vehicles in space and demonstrated great piloting skill in overcoming the thruster problem and bringing the spacecraft to a safe landing.

-more-

NAME: Russell L. Schweickart (Mr.)
       Lunar Module Pilot

BIRTHPLACE AND DATE: Born Oct. 25, 1935, in Neptune, N. J.
       His parents, Mr. and Mrs. George Schweickart, reside
       in Sea Girt, N.J.

PHYSICAL DESCRIPTION: Red hair; blue eyes; height: 6 feet;
       weight: 161 pounds.

EDUCATION: Graduated from Manasquan High School, N.J.; received
       a Bachelor of Science degree in Aeronautical Engineering
       and a Master of Science degree in Aeronautics and Astro-
       nautics from Massachusetts Institute of Technology.

MARITAL STATUS: Married to the former Clare G. Whitfield of
       Atlanta, Ga. Her parents are the Randolph Whitfields
       of Atlanta.

CHILDREN: Vicki, Sept. 12, 1959; Randolph and Russell, Sept. 8,
       1960; Elin, Oct. 19, 1961; Diana, July 26, 1964.

OTHER ACTIVITIES: His hobbies are amateur astronomy, photography
       and electronics.

ORGANIZATIONS: Member of the Sigma Xi.

EXPERIENCE: Schweickart served as a pilot in the United States
       Air Force and Air National Guard from 1956 to 1963.

       He was a research scientist at the Experimental Astronomy
       Laboratory at MIT and his work there included research
       in upper atmospheric physics, star tracking and stabili-
       zation of stellar images. His thesis for a Master's
       degree at MIT concerned stratospheric radiance.

       Of the 2,400 hours flight time he has logged, 2,100 hours
       are in jet aircraft.

       Schweickart was one of the third group of astronauts named
       by NASA in Oct. 1963.

-more-

APOLLO PROGRAM MANAGEMENT/CONTRACTORS

Direction of the Apollo Program, the United States' effort to land men on the Moon and return them safely to Earth before 1970, is the responsibility of the Office of Manned Space Flight (OMSF), National Aeronautics and Space Administration, Washington, D.C. Dr. George E. Mueller is Associate Administrator for Manned Space Flight.

NASA Manned Spacecraft Center (MSC), Houston, is responsible for development of the Apollo spacecraft, flight crew training and flight control. Dr. Robert R. Gilruth is Center Director.

NASA Marshall Space Flight Center (MSFC), Huntsville, Ala., is responsible for development of the Saturn launch vehicles. Dr. Wernher von Braun is Center Director.

NASA John F. Kennedy Space Center (KSC), Fla., is responsible for Apollo/Saturn launch operations. Dr. Kurt H. Debus is Center Director.

NASA Goddard Space Flight Center (GSFC), Greenbelt, Md., manages the Manned Space Flight Network under the direction of the NASA Office of Tracking and Data Acquisition (OTDA). Gerald M. Truszynski is Associate Administrator for Tracking and Data Acquisition. Dr. John F. Clark is Director of GSFC.

Apollo/Saturn Officials

NASA Headquarters

Lt. Gen. Sam C. Phillips, (USAF)	Apollo Program Director, OMSF
George H. Hage	Apollo Program Deputy Director, Mission Director, OMSF
Chester M. Lee	Assistant Mission Director, OMSF
Col. Thomas H. McMullen (USAF)	Assistant Mission Director, OMSF
Maj. Gen. James W. Humphreys, Jr.	Director of Space Medicine, OMSF
Worman Pozinsky	Director, Network Support Implementation Div., OTDA

Manned Spacecraft Center

George M. Low	Manager, Apollo Spacecraft Program
Kenneth S. Kleinknecht	Manager, Command and Service Modules
Brig. Gen. C. H. Bolender (USAF)	Manager, Lunar Module
Donald K. Slayton	Director of Flight Crew Operations
Christopher C. Kraft, Jr.	Director of Flight Operations
Eugene F. Kranz	Flight Director
Gerald Griffin	Flight Director
M. P. Frank	Flight Director
Charles A. Berry	Director of Medical Research and Operations

Marshall Space Flight Center

Maj. Gen. Edmund F. O'Connor	Director of Industrial Operations
Dr. F. A. Speer	Director of Mission Operations
Lee B. James	Manager, Saturn V Program Office
William D. Brown	Manager, Engine Program Office

Kennedy Space Center

Miles Ross	Deputy Director, Center Operations
Rear Adm. Roderick O. Middleton (USN)	Manager, Apollo Program Office
Rocco A. Petrone	Director, Launch Operations

Walter J. Kapryan — Deputy Director, Launch Operations

Dr. Hans F. Gruene — Director, Launch Vehicle Operations

John J. Williams — Director, Spacecraft Operations

Paul C. Donnelly — Launch Operations Manager

Goddard Space Flight Center

Ozro M. Covington — Assistant Director for Manned Space Flight Tracking

Henry F. Thompson — Deputy Assistant Director for Manned Space Flight Support

H. William Wood — Chief, Manned Flight Operations Div.

Tecwyn Roberts — Chief, Manned Flight Engineering Div.

Department of Defense

Maj. Gen. Vincent G. Huston, (USAF) — DOD Manager of Manned Space Flight Support Operations

Maj. Gen. David M. Jones, (USAF) — Deputy DOD Manager of Manned Space Flight Support Operations, Commander of USAF Eastern Test Range

Rear Adm. P. S. McManus, (USN) — Commander of Combined Task Force 140, Atlantic Recovery Area

Rear Adm. F. E. Bakutis, (USN) — Commander of Combined Task Force 130, Pacific Recovery Area

Col. Royce G. Olson, (USAF) — Director of DOD Manned Space Flight Office

Brig. Gen. Allison C. Brooks, (USAF) — Commander Aerospace Rescue and Recovery Service

Major Apollo/Saturn V Contractors

Contractor	Item
Bellcomm Washington, D.C.	Apollo Systems Engineering
The Boeing Co. Washington, D.C.	Technical Integration and Evaluation
General Electric-Apollo Support Dept., Daytona Beach, Fla.	Apollo Checkout and Reliability
North American Rockwell Corp. Space Div., Downey, Calif.	Spacecraft Command and Service Modules
Grumman Aircraft Engineering Corp., Bethpage, N.Y.	Lunar Module
Massachusetts Institute of Technology, Cambridge, Mass.	Guidance & Navigation (Technical Management)
General Motors Corp., AC Electronics Div., Milwaukee	Guidance & Navigation (Manufacturing)
TRW Systems Inc. Redondo Beach, Calif.	Trajectory Analysis
Avco Corp., Space Systems Div., Lowell, Mass.	Heat Shield Ablative Material
North American Rockwell Corp. Rocketdyne Div., Canoga Park, Calif.	J-2 Engines, F-1 Engines
The Boeing Co. New Orleans	First Stages (SIC) of Saturn V Flight Vehicles, Saturn V Systems Engineering and Integration Ground Support Equipment
North American Rockwell Corp. Space Div. Seal Beach, Calif.	Development and Production of Saturn V Second Stage (S-II)
McDonnell Douglas Astronautics Co. Huntington Beach, Calif.	Development and Production of Saturn V Third Stage (S-IVB)

-more-

International Business Machines
Federal Systems Div.
Huntsville, Ala.

Instrument Unit (Prime Contractor)

Bendix Corp.
Navigation and Control Div.
Teterboro, N. J.

Guidance Components for Instrument Unit (Including ST-124M Stabilized Platform)

Trans World Airlines, Inc.

Installation Support, KSC

Federal Electric Corp.

Communications and Instrumentation Support, KSC

Bendix Field Engineering Corp.

Launch Operations/Complex Support, KSC

Catalytic-Dow

Facilities Engineering and Modifications, KSC

ILC Industries
Dover, Del.

Space Suits

Radio Corp. of America
Van Nuys, Calif.

110A Computer - Saturn Checkout

Sanders Associates
Nashua, N. H.

Operational Display Systems Saturn

Brown Engineering
Huntsville, Ala.

Discrete Controls

Ingalls Iron Works
Birmingham, Ala.

Mobile Launchers (structural work)

Smith/Ernst (Joint Venture)
Tampa, Fla.
Washington, D.C.

Electrical Mechanical Portion of MLs

Power Shovel, Inc.
Marion, Ohio

Crawler-Transporter

Hayes International
Birmingham, Ala.

Mobile Launcher Service Arms

-more-

APOLLO 9 GLOSSARY

Ablating Materials--Special heat-dissipating materials on the surface of a spacecraft that can be sacrificed (carried away, vaporized) during reentry.

Abort--The unscheduled intentional termination of a mission prior to its completion.

Accelerometer--An instrument to sense accelerative forces and convert them into corresponding electrical quantities usually for controlling, measuring, indicating or recording purposes.

Adapter Skirt--A flange or extension of a stage or section that provides a ready means of fitting another stage or section to it.

Apogee--The point at which a Moon or artificial satellite in its orbit is farthest from Earth.

Attitude--The position of an aerospace vehicle as determined by the inclination of its axes to some frame of reference; for Apollo, an inertial, space-fixed reference is used.

Burnout--The point when combustion ceases in a rocket engine.

Canard--A short, stubby wing-like element affixed to the launch escape tower to provide CM blunt end forward aerodynamic capture during an abort.

Celestial Guidance--The guidance of a vehicle by reference to celestial bodies.

Celestial Mechanics--The science that deals primarily with the effect of force as an agent in determining the orbital paths of celestial bodies.

Closed Loop--Automatic control units linked together with a process to form an endless chain.

Deboost--A retrograde maneuver which lowers either perigee or apogee of an orbiting spacecraft. Not to be confused with deorbit.

Delta V--Velocity change.

Digital Computer--A computer in which quantities are represented numerically and which can be used to solve complex problems.

Down-Link--The part of a communication system that receives, processes and displays data from a spacecraft.

Ephemeris--Orbital measurements (apogee, perigee, inclination,
period, etc.) of one celestial body in relation to another
at given times. In spaceflight, the orbital measurements
of a spacecraft relative to the celestial body about which
it orbited.

Explosive Bolts--Bolts destroyed or severed by a surrounding
explosive charge which can be activated by an electrical
impulse.

Fairing--A piece, part or structure having a smooth, stream-
lined outline, used to cover a nonstreamlined object or
to smooth a junction.

Flight Control System--A system that serves to maintain attitude
stability and control during flight.

Fuel Cell--An electrochemical generator in which the chemical
energy from the reaction of oxygen and a fuel is converted
directly into electricity.

G or G Force--Force exerted upon an object by gravity or by
reaction to acceleration or deceleration, as in a change
of direction: one G is the measure of the gravitational
pull required to accelerate a body at the rate of about
32.16 feet-per-second.

Gimballed Motor--A rocket motor mounted on gimbal; i.e., on a
contrivance having two mutually perpendicular axes of rotation, so as to obtain pitching and yawing correction
moments.

Guidance System--A system which measures and evaluates flight
information, correlates this with target data, converts
the result into the conditions necessary to achieve the
desired flight path, and communicates this data in the form
of commands to the flight control system.

Inertial Guidance--Guidance by means of the measurement and
integration of acceleration from onboard the spacecraft.
A sophisticated automatic navigation system using gyroscopic
devices, accelerometers etc., for high-speed vehicles. It
absorbs and interprets such data as speed, position, etc.,
and automatically adjusts the vehicle to a pre-determined
flight path. Essentially, it knows where it's going and
where it is by knowing where it came from and how it got
there. It does not give out any radio frequency signal so
it cannot be detected by radar or jammed.

Injection--The process of boosting a spacecraft into a calcu-
lated trajectory.

Insertion--The process of boosting a spacecraft into an orbit around the Earth or other celestial bodies.

Multiplexing--The simultaneous transmission of two or more signals within a single channel. The three basic methods of multiplexing involve the separation of signals by time division, frequency division and phase division.

Optical Navigation--Navigation by sight, as opposed to inertial methods, using stars or other visible objects as reference.

Oxidizer--In a rocket propellant, a substance such as liquid oxygen or nitrogen tetroxide which supports combustion of the fuel.

Penumbra--Semi-dark portion of a shadow in which light is partly cut off, e.g., surface of Moon or Earth away from Sun. (See umbra.)

Perigee--Point at which a Moon or an artifical satellite in its orbit is closest to the Earth.

Pitch--The angular displacement of a space vehicle about its lateral axis (Y).

Reentry--The return of a spacecraft that reenters the atmosphere after flight above it.

Retrorocket--A rocket that gives thrust in a direction opposite to the direction of the object's motion.

Roll--The angular displacement of a space vehicle about its longitudinal (X) axis.

S-Band--A radio-frequency band of 1,550 to 5,200 megahertz.

Sidereal--Adjective relating to measurement of time, position or angle in relation to the celestial sphere and the vernal equinox.

State vector--Ground-generated spacecraft position, velocity and timing information uplinked to the spacecraft computer for crew use as a navigational reference.

Telemetering--A system for taking measurements within an aero-space vehicle in flight and transmitting them by radio to a ground station.

Terminator--Separation line between lighted and dark portions of celestial body which is not self luminous.

Ullage--The volume in a closed tank or container that is not occupied by the stored liquid; the ratio of this volume to the total volume of the tank; also an acceleration to force propellants into the engine pump intake lines before ignition.

Umbra--Darkest part of a shadow in which light is completely absent, e.g., surface of Moon or Earth away from Sun.

Update pad--Information on spacecraft attitudes, thrust values, event times, navigational data, etc., voiced up to the crew in standard formats according to the purpose, e.g., maneuver update, navigation check, landmark tracking, entry update, etc.

Up-Link Data--Information fed by radio signal from the ground to a spacecraft.

Yaw--Angular displacement of a space vehicle about its vertical (Z) axis.

-more-

APOLLO 9 ACRONYMS AND ABBREVIATIONS

(Note: This list makes no attempt to include all Apollo program acronyms and abbreviations, but several are listed that will be encountered frequently in the Apollo 9 mission. Where pronounced as words in air-to-ground transmissions, acronyms are phonetically shown in parentheses. Otherwise, abbreviations are sounded out by letter.)

AGS	(Aggs)	Abort Guidance System (LM)
AK		Apogee kick
APS	(Apps)	Ascent Propulsion System
BMAG	(Bee-mag)	Body mounted attitude gyro
CDH		Constant delta height
CMC		Command Module Computer
COI		Contingency orbit insertion
CSI		Concentric sequence initiate
DAP	(Dapp)	Digital autopilot
DEDA	(Dee-da)	Data Entry and Display Assembly (LM AGS)
DFI		Development flight instrumentation
DPS	(Dips)	Descent propulsion system
DSKY	(Diskey)	Display and keyboard
FDAI		Flight director attitude indicator
FITH	(Fith)	Fire in the hole (LM ascent staging)
FTP		Fixed throttle point
HGA		High-gain antenna
IMU		Inertial measurement unit

- more -

IRIG	(Ear-ig)	Inertial rate integrating gyro
MCC		Mission Control Center
MC&W		Master caution and warning
MTVC		Manual thrust vector control
NCC		Combined corrective maneuver
NSR		Coelliptical maneuver
PIPA	(Pippa)	Pulse integrating pendulous accelerometer
PLSS	(Pliss)	Portable life support system
PUGS	(Pugs)	Propellant utilization and gaging system
REFSMMAT	(Refsmat)	Reference to stable member matrix
RHC		Rotation hand controller
RTC		Real-time command
SCS		Stabilization and control system
SLA	(Slah)	Spacecraft LM adapter
SPS		Service propulsion system
THC		Thrust hand controller
TPF		Terminal phase finalization
TPI		Terminal phase initiate
TVC		Thrust vector control

CONVERSION FACTORS

	Multiply	By	To Obtain
Distance:	feet	0.3048	meters
	meters	3.281	feet
	kilometers	3281	feet
	kilometers	0.6214	statute miles
	statute miles	1.609	kilometers
	nautical miles	1.852	kilometers
	nautical miles	1.1508	statute miles
	statute miles	0.86898	nautical miles
	statute mile	1760	yards
Velocity:	feet/sec	0.3048	meters/sec
	meters/sec	3.281	feet/sec
	meters/sec	2.237	statute mph
	feet/sec	0.6818	statute miles/hr
	feet/sec	0.5925	nautical miles/hr
	statute miles/hr	1.609	km/hr
	nautical miles/hr (knots)	1.852	km/hr
	km/hr	0.6214	statute miles/hr
Liquid measure, weight:	gallons	3.785	liters
	liters	0.2642	gallons
	pounds	0.4536	kilograms
	kilograms	2.205	pounds

- more -

	Multiply	By	To Obtain
Volume:			
	cubic feet	0.02832	cubic meters
Pressure:			
	pounds/sq inch	70.31	grams/sq cm

-end-

www.ingramcontent.com/pod-product-compliance
Lightning Source LLC
Chambersburg PA
CBHW051217200326
41519CB00025B/7154

* 9 7 8 1 7 8 0 3 9 8 5 8 7 *